W9-CEW-305

Ecological Identity

Ecological Identity

Becoming a Reflective Environmentalist

Mitchell Thomashow

The MIT Press
Cambridge, Massachusetts
London England

This book was set in Palatino by The MIT Press and printed and bound in the United States of America.

Library of Congress Cataloging-in-Publication Data

Thomashow, Mitchell.
Ecological identity: becoming a reflective environmentalist / Mitchell Thomashow.
p. cm.
Includes bibliographical references and index.
ISBN 0-262-20100-3
1. Environmentalism. 2. Green movement. 3. Environmental education. 4. Environmental policy—Citizen participation.
I. Title.
GE195.T48 1995
363.7'05—DC20 94-36471
CIP

"What I propose, therefore, is very simple: it is nothing more than to think what we are doing."

Hannah Arendt, *The Human Condition*

Contents

List of Learning Activities

Introduction: Becoming a Reflective Environmentalist

Finding the Common Trailhead

Twenty different people. Twenty converging paths. An instant group, but not yet a learning community. The faces that enter this room reflect curiosity, trepidation, and anticipation: yearnings to explore how they arrived at this common trailhead. They wonder what they are walking into.

John is a hazardous waste technician. He travels around the Northeast cleaning up toxic waste dumps. Sloshing around in the industrial muck, decked out in the full regalia of protective clothing, he is engaged in the grunt work of environmental pollution. Recently, as he was cleaning up a toxic waste site, three children looked at him from behind a fence, wondering when they could play in their field again. John wouldn't be here otherwise.

As a selectwoman in a small New England town, Betsy thinks she has seen it all. She understands the pressures of development and the importance of bringing new jobs to the town, but she believes strongly in the need for open space and conservation. She assumes that she has balanced these concerns well, but one day a challenging proposal crosses her desk. It threatens the integrity of her own special place, the land she walks for solace and inspiration.

Steve is a political activist who has spent 20 years working in grassroots communities, listening to people voice their concerns about the health of their families, the safety of their workplaces, and the security of their jobs. While backpacking in the wilderness, he realizes that his passion for the earth is intricately connected to his political work in the city.

The chairs are arranged in a circle, wide enough for twenty participants. As I arrive, I observe that I am subtly, but carefully scrutinized. There are some familiar faces, but many of these people are new to me, students just starting their graduate program, who haven't been in school for many years. They glance at one another, wondering who these new colleagues are, what they have in common, and what they might learn from their work together. Undoubtedly they've been told by other students or by faculty advisors that my class will encourage them to look carefully at their environmental values, yet they are not sure exactly what that means or whether it is even something they want to do.

I teach a graduate class called Patterns of Environmentalism. Despite having taught this course for 15 years, I always grow somewhat anxious before the first meeting. I sense the excitement that pervades the students. I consider the extraordinary journey we will embark on together as our collective experiences unfold and we become immersed in reflection about the meaning of environmentalism. I know that I will uncover a range of emotional and intellectual impressions and that all of us will emerge from the class changed in some profound way. It happens this way every semester.

I ask the class members to introduce themselves, encourage them to tell about the paths that led them here, why they are in environmental studies, what they expect from my class, and what they hope to attain. I am always impressed by the diversity of their responses.

Consider a typical class profile. There are enivronmental educators who work at urban or rural nature centers, sanctuaries, parks, and schools. There are policy people who work for the Environmental Protection Agency (EPA), regional planning agencies, think tanks, advocacy organizations, private consulting firms, state and local government. There are people from the communications field, who are writers, interpreters, museum designers, or who may be involved in public relations. And there are the technical professionals, who work with pollution control industries, or are involved with hazardous waste cleanup, water quality, or in any of the spiraling environmental science professions. New occupational niches have developed: people have found roles in organizations and have job descriptions that were not even conceivable several years ago.

But there are also those new to the environmental profession, abandoning their former careers, making the difficult and risky decision to try something different, pursuing a professional direction reflecting their powerful environmental concerns. These include laboratory technicians, computer programmers, social workers, homemakers, carpenters, health professionals, and lawyers. Sometimes there are people still searching for appropriate career paths, thinking they may have found one in environmentalism. These people demonstrate courage in redefining their professional goals and aspirations.

The list of personal expectations for the class is equally long. Some people cite their interest in learning more about environmentalism, hoping this will provide a context for their own work. Others put it in more psychological terms, expressing the need to understand their deepest motivations and values regarding environmental issues, desiring to probe their experiences and feelings and make more sense out of them, wondering how they can more directly experience nature and apply those experiences to their personal and professional decisions. And there are some who aren't quite sure what to say, not yet able to articulate the psychological and philosophical aspects of their environmental experience.

After the introductions are completed, I pause, allowing the class to digest all they have just heard. I observe that our small group is in many ways a microcosm of the environmental profession. Despite our diverse experiences and backgrounds, we are called to environmentalism not only because we want to make the world a better place or because we are interested in the subject matter but because the ideas of environmentalism speak to something deep inside us. I reveal to the class that I am here for the same reasons they are. I want to explore why I consider myself an environmentalist, what this means in terms of my personal and professional choices, how I use my environmental values to construct a personal identity. This class is a phase of a lifelong search to understand my place in the ecosystem, my role and purpose as a human being. I need a strong and cohesive learning community to help me formulate these ideas. We all need one another for this purpose. The circle is temporarily quiet as we gaze around. As I look at the class, as the class looks at me, we realize that we are all searching together. We see it in one another's eyes. We are becoming reflective environmentalists, and we will use our personal experiences and our interpretation of the great texts of environmentalism as a means for common reflection.

For most of my career, I have been teaching courses and facilitating workshops about what attracts people to environmentalism, listening carefully to the interests and experiences of a broad cross section of environmentally concerned people. Echoes of nourishing and challenging discussions reverberate as I watch yet another group begin its journey. There was the woman in Prague who bemoaned the devastating environmental pollution of her home town in Bohemia. The Arabic Israeli who recalled how every spring flowers bloomed on the thatched roof of his house and how his family made an annual barefoot trek to the shores of the Mediterranean. The busy executive director of a thriving environmental organization who wished she could spend more time in the mountains to nourish her soul, to appreciate the magnificent sensory environment. The environmental educator from California whose face came alive as he described what it is like teaching natural history to young children.

But you don't have to be a professional environmentalist, or leave your job, or experience a life transition to have similar experiences. It is not just people in my class or students in environmental studies programs who have a desire to talk about these issues. There are hundreds and thousands of men, women, and children who have a compelling urge to reincorporate nature in their lives, to have what they consider a direct experience of nature, and wonder how such direct experiences lend meaning to their lives. Whether it's a birder in Central Park, a business person on a rafting trip, a child growing tomatoes in an urban garden, a computer programmer who enjoys backpacking, a lone fisherman on a northern lake—there are many expressions of this common impulse.

Equally important is the perception of the threats to nature, which may be revealed in many different ways: the community member who is concerned about unplanned development, the parent who has been informed of asbestos in the local schools, the working person who is concerned about environmental hazards on the job, the student who learned about ozone depletion in her high school science class.

Preparing to Hike the Trail

In this book, I reveal the depth of personal expression at the core of contemporary environmentalism. Many people experience an inner voice that compels them to explore their relationship to the natural world. Whether it is the joy and inspiration that accompanies a backpacking experience, or the sense of loss and despair about an endangered habitat, these are common experiences that not only motivate people to take action regarding environmental issues, to volunteer for environmental organizations and to choose to enter the environmental profession but often cause them to think carefully about the moral foundations of their everyday decisions. Indeed, these experiences of nature become a foundation for the merging of the secular and the sacred, using the metaphors of ecology and environment. They constitute a common vision, an attempt to define the moral parameters for an ecologically sound way of life, and to use this vision to provide wisdom and guidance.

This is the common trailhead, the place where many paths converge. To hike the path implies a searching process, a search to recover and reclaim the importance of nature in one's personal development. I suggest that this orientation and sensibility involves a reconstruction of personal identity, so that people begin to consider how their actions, values, and ideals are framed according to their perceptions of nature. This is the basis of what I call ecological identity work: using the direct experience of nature as a framework for personal decisions, professional choices, political action, and spiritual inquiry. Ecological identity refers to how people perceive themselves in reference to nature, as living and breathing beings connected to the rhythms of the earth, the biogeochemical cycles, the grand and complex diversity of ecological systems.

In my classes, and in this book, I provide a guide to the learning experiences that accompany ecological identity work. This represents a profound educational challenge for anyone who is interested in environmental issues: the necessity of translating the inner voice that yearns to speak with nature into a broad understanding of ecology, community, and citizenship. When the inner voice begins to speak, a learning space emerges, an opportunity for the cultivation of awareness. This book is about how to explore, interpret, and nurture these learning spaces, the openings made possible when people are moved to contemplate their experience of nature. My perspective is that this is most effectively accomplished when a person learns how to reflect on, discuss, and ultimately internalize the personal and public impact of environmental experiences, and to do so within both private and public life, in classrooms and town meetings, on street corners and in living rooms, on the trail or at the dump.

Hence my approach is both practical and theoretical. *Ecological Identity* is sprinkled with activities and anecdotes, the real stories of everyday life, and the challenges and frustrations that accompany any new learning experience. It is a source book for personal and collective reflection about environmental issues.

There are five perspectives or voices that characterize the flow of the book. Here is a description of what you can expect. I write about my experiences working with adult learners, graduate students who are or aspire to become environmental professionals, a broad array of people who enter the environmental profession because of a strong inner drive, even a calling, to understand the meaning of their lives, to contribute what they can to promote environmental quality, and to claim a moral basis for their professional work and personal choices. Most of my career has been spent designing programs, courses, and curriculum to meet the needs of environmental professionals. I have found that the most critical component of their training is the ability and willingness to look deeply within themselves, to understand their motivations and aspirations, to clearly articulate their environmental values, and know how to apply them to professional and personal decisions. Through a series of experiential learning approaches, I have become engaged with my students in the process of confronting these issues. These activities constitute an emerging text, a collaborative voice, a powerful means to bridge the inner and outer life of the environmentalist. I include many excerpts from student work: these are the stories of their experiences, the voices of ecological identity.

I juxtapose these activities with stories of my own experience, and the ways in which ecological identity work permeates my life. I have chosen ordinary circumstances, the kinds of things that happen to most people: discussions with neighbors, experiences in the outdoors, thoughts I have while driving to work, the dilemmas I face navigating the demands and responsibilities of work or parenting. These are the quotidian events and circumstances that I ordinarily take for granted. What happens when the lid is opened, when these experiences are subjected to reflection and analysis, and the political and ecological consequences of everyday actions are the subject of intense scrutiny? These personal stories make up the curriculum of ecological identity, the laboratory of learning.

I introduce a scholarly interpretation, exploring how the collaborative voice of my students and my personal stories have a theoretical context. I describe how ecological identity is intrinsic to contemporary environmentalism, how it is a means through which to interpret the history of American environmentalism, an approach to carving out an environmental citizenship, and an orientation for how environmental studies might be taught. I weave a synergy of scholarship, a collage of collective stories and theoretical inquiry. This appears recursive at times, as various themes unfold and reiterate throughout the book. The scholar in me delights in creating order and meaning, but this material is complex, and it rearranges itself in many different ways. I find that my "book teachers," Henry David Thoreau, Gary Snyder, Joanna Macy, David Orr, Thich Nhat Hanh, Benjamin Barber, and many others pop up repeatedly, helping to create a collage of interpretation.

I highlight many of the tensions, contradictions, uncertainties, and ambiguities of constructing an ecological identity in the shifting terrain of post-modern life. It is not easy to navigate this terrain, balancing conflicting feelings about technology, politics, faith, nature, and humans. In this book, I try to bring these tensions to the surface by naming and confronting them. There are no easy answers, comprehensive programs, or miraculous blueprints that will guide us smoothly to an ecotopian future. Becoming a reflective environmentalist brings happiness and struggle, liberation and suffering. Many of the questions I raise remain unresolved. My strategy is to provide the learning tools that enable people to interpret these difficult questions, to bring them to the surface of their daily lives, encourage public discussion about them, emphasize their educational importance, and uncover the elegance of uncertainty juxtaposed with the strength of clarity and moral resolve.

This book revolves around learning activities that are designed to stimulate reflective, introspective analysis. I include these in the spirit of participatory learning; to encourage mindfulness, group discussion, and public controversy. Although I have used these activities over many years with adult learners in a graduate school setting, my students have applied variations in other educational and workplace environments: elementary schools, nature trails, public forums, organizational retreats, family rituals, and so on. Certainly I invite the professional educator to put these activities to good use, but I encourage all readers to consider using them, either in a public setting, or as a supplement to reading the book. In this way, the book becomes *your* text as well as mine; a hypertext, if you will, in which the full dimensions of your personal experience are linked to a unique, active, participatory interpretation.

Reading the Topographic Map

These perspectives or voices characterize the flow of the book. In a sense, the book is hologrammatic. From any particular section, the entire book may unfold. Of course, there are chapters and sections which serve to delineate the primary themes, but often I work with several ideas simultaneously, and throughout the book they look somewhat different, depending on the themes and the voice. I have arranged a linear sequence, and like most books, it moves from point *A* to point *B*. But I'm well aware that the busy, postmodern reader will often scan a book, dipping in and out to find a unique path through the text. Each chapter has been designed as a self-contained unit, and although there is a logic behind the flow of the book, the interested reader can find many places to begin.

In chapter 1, I explore some of the different ways that people describe their experiences of nature, the means of personal expression that constitute the voice of ecological identity. Memories of childhood places, the perception of disturbed places, the contemplation of wild places: these are examples of transformational moments in people's lives, when they realize that their personal identity is intrinsically connected to their direct experience of nature. Using these examples as a guide, I consider the significance and definition of ecological identity as an integrative concept and discuss some of its educational implications.

Chapter 2 describes environmentalism as an emerging interpretive tradition. I show how Henry David Thoreau, John Muir, and Rachel Carson serve as environmental archetypes, and how they are important as ecological identity role models. How do their experiences, dilemmas, and struggles inform the work of contemporary environmentalists? If Thoreau, Muir, and Carson represent the roots of modern environmentalism, they are the foundation of a large tree, comprising many branches, reflecting diverse, but integrated points of view. I describe an educational process in which people construct artistic representations of trees as a means of exploring the contemporary environmental spectrum, using the trees to search the movements and organizations of the environmental field. This suggests a process of personal and collective reflection in which people create individualized interpretations of environmentalism as a means of ecological identity work.

The public expression of ecological identity leads to involvement in the political arena and immersion in the controversies that surround environmental issues. Yet many people are ambivalent about political expression, and there are good reasons why people avoid involvement in public life. Chapters 3 and 4 describe the intricate relationship between environmentalism and citizenship. Using learning activities such as *the property list, the community network map, political autobiography,* and *power flow analysis,* these chapters explore the integration of ecological and political identity, how to forge a concept of ecological citizenship, and how to formulate a political voice for environmental issues.

Throughout the book there is an implicit assumption that ecological identity work represents a process of personal and global healing. Environmentalists carry a profound dual burden: how can they at once convey a sense of wonder and appreciation about the natural world, and also be the harbingers of impending doom, warning the world about ecological catastrophes? In chapter 5 I suggest that this extraordinary psychological and philosophical challenge is a critical agenda for environmental organizations and is the basis of *reflective environmental practice.*

Chapter 6 describes the relationship between ecological identity work and reflective learning generally. In so doing, it serves as an educational rationale for this book. This chapter includes the design criteria for effective learning activities, what I call *interpretive modalities,* providing a short handbook for the educator and the concerned citizen: how do you develop vibrant and vital learning communities for ecological identity work? I suggest that four basic questions should be at the heart of environmental education: What do I know about the place where I live? Where do things come from? How do I connect to the earth? What is my purpose as a human being? I integrate these questions around the *sense-of-place map,* showing that ecological identity work belongs in the school, the home, the community, and the workplace.

Who Should Take This Hike?

Ecological Identity is intended for several audiences. First, it is designed for working environmentalists who wish to look more deeply at the context of their work, where it lies in the broad spectrum of environmental thinking, how they might expand their concept of environmentalism so it becomes a bridge between professional practice, personal lifestyle, living in nature, and political participation. Second, it should be useful for the person who is new to environmentalism, who is interested in some of the questions that environmentalism raises and would like to look at them more carefully. In this way the book serves as an introduction to contemporary environmentalism through the lens of ecological identity. Finally, it presents a challenge for the academic environmentalist, introducing ecological identity work as an educational basis for the teaching of environmental studies.

In writing this book, I feel as if I am exploring a rocky coastline. The material is a fractal; at every stage and scale, the shapes and structures form an inclusive pattern, but the landscape never looks quite the same. Each nook and cranny reveals more detail, more possibilities requiring additional exploration—a coastline is infinite. Each chapter becomes a universe of learning. By covering such a wide range of material, I inevitably forget some important dimension, I miss an important view, I trip and stumble on a slippery slope.

Yet this is the coastline that I choose to travel. I approach the journey with humility, always surprised at where it leads, respectful of how much I fail to see and of what remains uncovered. The most profound learning occurs through the process of the search itself. My wish is that this book serve as a tool for exploration, a handbook to the infinite coastline, a way of becoming a reflective environmentalist.

Acknowledgments

I sit on the soft forest floor in a bed of pine needles, my back resting against a small glacial boulder, listening to the mild southern breeze on this gentle September morning. As I wrote *Ecological Identity,* I would alternate between this spot and my study, a writing pad and the computer, trying to balance sensory awareness, personal reflection, and detached intellect. I recall the tensions, struggles, and pleasures of the writing process—how this place, nestled in the foothills of Mount Monadnock in southwest New Hampshire, provided nourishment and perspective.

I feel waves of gratitude and respect as I consider the remarkable gifts of support and critique that accompanied this extraordinary learning experience. Images of current and former students appear, their enthusiasm and engagement, their hopes and joys, their remarkable creativity. This is your book. I hope I represent the breadth of your insight and the depth of your concerns.

I thank the many manuscript readers, especially those who took the time to offer written comments: John Tallmadge, Steve Chase, Kathleen Hogan, John Elder, Larry Morris, Helmut Schreier, Stephanie Kaza, Susan Clark, Jimmy Karlan, Robert D. Kahn, Alexandra Dawson, David Singer, Sabine Hrachdekian, and Katie Hennessey. Thank you, Sy Montgomery, for all your support, humor, and encouragement. The anonymous reviewers, chosen by the MIT Press, had extremely constructive suggestions.

At a small graduate school where most faculty share academic and administrative responsibility, it is difficult to take time away from the pressing demands of programs and courses. Inevitably other people have carried the burden of my periodic absence or temporary preoccupation. With great warmth I thank the faculty and staff of the Environmental Studies Department of Antioch New England

Graduate School, particularly Maich Gardner, a superb administrator and sensitive colleague. Special thanks to Provost Jim Craiglow, who understands diverse approaches to scholarship and lets people actualize their visions. Thank you, too, Ty Minton, for taking time away from your impossible schedule to help me with the charts. I wish to thank Lawrence Goldsmith for the use of his beautiful painting and Fran Silvestri for his help in selecting and photographing it.

The MIT Press has been a responsive and supportive publisher. This project was seeded during a discussion I had with Frank Urbanowski as we drove to the Los Angeles airport, returning from an environmentalist retreat. In his understated, solid way he encouraged this book at all stages of its development. Madeline Sunley has helped me place the book in perspective, urging clarity, counseling patience, offering balance and vision. H. Emerson (Chip) Blake of *Orion* magazine is a superb editor—ubiquitous and invisible, demanding and patient, direct and open-minded. Chip provided professional and personal support at the most crucial times.

I offer deep gratitude to my family. Jessica and Jacob, my children, for respecting my space and sharing your magic. Milton and Elissa, my parents, for your sincere interest. And my colleague and wife, Cindy Thomashow, for your unstinting support and encouragement. I would follow Cindy around the house, call her at work, interrupt her at inopportune times, urging her to listen to a "short" passage. She would always attentively comply, Cindy understands the intellectual and emotional swings of the writing process—she heard every passage and idea in the book repeatedly, providing outstanding suggestions, integrating her vision and common sense as an environmental educator. She knew when to let me work and how to pull me away. Her love, selflessness, and strength are a beacon.

My attention returns to the forest. A warm sun is gleaming in the yellow-green September light. Writing acknowledgments is a delightful closure—a reminder that ecological identity work involves people and nature, learning how to live in a place, taking pleasure in the gifts of life and mind. I am compelled to take a long walk, to celebrate this glorious autumn day, to move unencumbered through the glowing northern forest.

Dublin, New Hampshire, Monadnock Region
September 1994

Ecological Identity

1 Voices of Ecological Identity

A Cluster of Concepts

If you want to stir up a rousing discussion among a diverse group of environmentalists, ask them what they think environmentalism means. During the first session of my Patterns of Environmentalism class, this is exactly what I do.

I use a technique called the cluster diagram. You take a sheet of newsprint and write a word, in this case *environmentalism,* in the middle of the paper and construct a free-flowing diagram, letting your thoughts go wherever they may. Roaming around the room, I watch twenty adults on their hands and knees scribbling their ideas, synchronized with the squeak of magic markers. When they complete their diagrams, they place them around the walls of the classroom, so the entire group can reflect on all the places the word environmentalism has taken them.

What a conglomeration of words and ideas! The class surveys the motley assortment of semiorganized scribbles, collecting their impressions, perhaps slightly bewildered at the scope of responses. The students return slowly to their chairs, and we begin our discussion. I ask them if they can detect any conceptual patterns, ways to organize this expansive material. Several distinct patterns emerge. It appears there are four coherent clusters of words, reflecting specific, but connected trains of thought.

For example, there are the words we typically associate with the environmental sciences: biodiversity, systems, biogeochemical cycles, biosphere, habitat, global change, pollution, extinction, and so on. Clearly environmentalism is informed by the power of these concepts. Each word suggests a series of complex relationships, implying a sophisticated cognitive understanding. And if we so choose, we can divide this cluster into more specific arrangements, reflecting the branches of environmental science and ecology.

Another cluster reflects the broad spectrum of contemporary environmental thought. These are the people, organizations, and ideas that influence the way we think about nature: Rachel Carson, David Brower, Greenpeace, conservation, deep ecology, ecofeminism, bioregionalism, sustainability, and so forth. And these names spark some colorful associations in their own right. They are sources of inspiration and controversy.

We form a third cluster from the long list of recent environmental catastrophes: Chernobyl, Three Mile Island, Bhopal, *Exxon Valdez.* These events leave a distinct symbolic impression, grim reminders of the human and animal suffering that accompanies environmental pollution. This list too can be broken down into a long chain of events, local or global, that have served as catalysts for environmental action and public concern.

Finally, there is a more complex cluster that seems to connect all of the others. These are the big words. They express how people perceive themselves in relationship to the earth: interconnected, interdependent, reverent, committed, vulnerable, compassionate. Words like these appear on almost all of the diagrams. Environmentalism connotes concepts *and* feelings. The class agrees that we hear these words a lot in relationship to feelings about the earth, so much, in fact, that they can easily lose their meaning. Yet surely there is a great depth of expression represented here, and perhaps by unraveling the visceral experiences they connote, we can begin to explore the deep values that attract people to environmentalism.

Concepts and feelings—this is the interface I want to probe. There is a dynamic relationship between the profound intellectual concepts of environmentalism and the memories and life experiences which validate them. My purpose in the class and my task in this chapter are one and the same: to show how an ecological worldview can be used to interpret personal experience, and how that interpretation leads to new ways of understanding personal identity. I call this process *ecological identity work.*

What Is Ecological Identity?

Ecological identity. What unique synergy results from joining these words into a unified concept? *The Oxford Dictionary of English*

Etymology defines *identity* as the quality of being the same.[1] That seems simple enough. The stickler is the object. Being the same as what or whom? Of course, from the perspective of modern social science, identity is a very complex notion, referring to all the different ways people construe themselves in social relationships as manifested in personality, values, actions, and sense of self. To have an identity crisis is to be lost in the world, lacking the ability (temporarily, one hopes) to connect the self to meaningful objects, people, or ideas—the typical sources of identification.

In *The Diversity of Life*, Edward O. Wilson defines *ecology* as "the scientific study of the interaction of organisms with their environment, including the physical environment and other organisms living in it."[2] That is about as broad as you can get, implying nothing less than a way to organize knowledge about nature. Ecology is used widely, not only as a scientific concept but also metaphorically to describe how humans interact with nature, what is often referred to as an *ecological worldview*. The prefix *eco-* is ubiquitous in the environmental literature, employed as a way to overlay a concept with an ecological connotation.

Adding the word *ecological* substantially challenges the notion of identity. Ecological identity refers to all the different ways people construe themselves in relationship to the earth as manifested in personality, values, actions, and sense of self. Nature becomes an object of identification. For the individual, this has extraordinary conceptual ramifications. The interpretation of life experience transcends social and cultural interactions. It also includes a person's connection to the earth, perception of the ecosystem, and direct experience of nature.

What does it mean to say that nature becomes an object of identification? This can entail considerable ambiguity. After all, the nature we are referring to is a social construction, a human concept, varying from culture to culture and person to person. Nature may refer to many things—the stars and the galaxies, the earth and its atmosphere, evolutionary and ecological processes, chipmunks and butterflies. Perhaps, it is safest to say that ecological identity describes how we extend our sense of self in relationship to nature, and that the degree of and objects of identification must be resolved individually. To be more specific, each person's path to ecological identity reflects his or her cognitive, intuitive, and affective perceptions of ecological relationships.[3]

Words such as *ecology, nature,* and *earth* are often used interchangeably, referring metaphorically to an ecological worldview. It is useful to recognize the colloquial freedom here, without pretending to arrive at any analytical precision. Suffice it to say, our present task is to understand the concepts and feelings attached to these words and how they might be used to reconstruct personal identity.

In a provocative paper, "Ecology and Identity," psychologist Richard Borden reviewed a program of research suggesting that the "study of ecology leads to changes of identity and psychological perspective, and can provide the foundations for an 'ecological identity'—a reframing of a person's point of view which restructures values, reorganizes perceptions and alters the individual's self-directed, social, and environmentally directed actions."[4] Borden and his colleagues were interested in the personality characteristics of people who demonstrated a high degree of ecological concern, knowledge, and action.

Characteristics that emerged among ecologically oriented people included: strong value orientation and ethical conscientiousness, relative freedom from self-preoccupation, cooperativeness, and leadership potential. Borden describes these people as showing "clear signs of being philosophically inclined, cultural-doubting, personalities who were arriving at unique solutions to their own identities with root metaphors that reflected their ideas of ecology."[5] Further, he found that the cognitive and emotional aspects of ecological attitudes are not necessarily related. In other words, either facts or feelings, depending on the individual and the circumstances, may precipitate reflection or action.

It is important not to overstate studies such as these. Although they yield interesting data about human personality, there are so many cultural and historical variables at work that we should be wary of drawing causative inferences. Rather, I view this information and other data from environmental psychology[6] as a way to support an interpretive orientation. Ecological identity has conceptual integrity because there is evidence suggesting that people take action, or formulate their personality based on their ecological worldview. Either a cognitive or intuitive understanding of ecology may significantly reorient personal identity.

Ecological Identity Work

As an educator, I am interested in setting up learning situations that allow people to reflect on the *possibilities* of ecological identity. The knowledge and experiences that constitute an ecological worldview can be used reflectively to reinterpret the memories, events, and circumstances of personal development. This is what I mean by 'ecological identity work,' the reflective processes we use to explore this realm. The educational relationships are fourfold: how people learn about ecology, how people perceive themselves in relationship to ecosystems, how an understanding of ecology changes the way people learn about themselves, and how an ecological worldview promotes personal change.

Why is ecological identity work important? It is the personal introspection that drives one's commitment to environmentalism. Environmentalism refers to the unfolding, evolving, active development of an ecological worldview, a perspective that is at once dynamic, diverse, and radical. It represents the ideas, people, and actions that constitute a social and intellectual movement. Ecological worldview is the constellation of words and images depicted in the students' cluster diagrams, an emerging philosophy, built on an intuitive and cognitive understanding of ecology, a way of interpreting the world, as applied to personal choices and public policy.[7]

Environmentalism is dynamic in the sense that it changes according to historical circumstances—the context of world events and the intellectual development of ecological thought. Its diversity is reflected in the great variety of intellectual approaches it entails (which is the subject of chapter 2). It is radical in that it challenges the way people view the world, presenting a moral and even spiritual approach that has significant ramifications for how people live their lives and conduct their affairs.

When environmentalists challenge transnational corporations in the Pacific Northwest, proclaiming that endangered species are more important than short-term economic growth, they are presenting a dramatic and threatening demand. When a community decides to establish strict zoning laws to protect open space in their efforts to maintain the ecological integrity of a region, they are probably doing so despite some intense pressures to develop the land. When a family decides that their way of life is profligate and wasteful and that they will participate in recycling, green consumerism, and contribute money and time to environmental organizations, they are involved in an important life change. In all of these cases, people are making decisions, or challenging public policy, and they are doing so based on their ecological worldview.

These are the situations faced by professional environmentalists and concerned citizens. They are typically controversial, sometimes ambiguous, and often filled with uncertainty. It is not always possible to distinguish right from wrong. And the most appropriate paths of action may involve significant personal risk. This risk comes not only from public controversy and personal sacrifice; it also represents the process of challenging one's own values, that is, being willing to ask what it means to look at life a little bit differently, in this case, from an ecological perspective. This reflective process can go much, much deeper. For some, an ecological worldview implies personal transformation as well.

My experience is with environmental practitioners. It is clear that when they enter the environmental profession, they are making a profound life choice. Somewhere in the career decision-making process, these people have been attracted to a livelihood in which they will be working, as they often say themselves, "to protect the environment." There is something within them that yearns for a more fulfilling relationship to the natural world and they seek to incorporate various aspects of their lives in the natural world in order to achieve that goal. In some cases, this is an image or an idea; it is romantic and intangible, based perhaps on a recent life circumstance, awakened by dissatisfaction or the intense desire to get something more out of life. For others, it is the obvious continuation of a lifelong pursuit. Most aspiring environmentalists perceive themselves as choosing more than a profession; they are searching to link their ecological worldview to their personal identity.

These people are all looking to cultivate their affiliations, to understand more about the American environmental tradition, and to understand their place in the environmental movement. They are looking for ways to support their environmentalism, often feeling frustrated that they are moving against a strong tide, and that their positions are not widely supported by the mainstream culture. So they search for ways to understand the core values that inspire their decisions and how to translate their environmental values to the culture at large, to make their ideas accessible and reasonable, to convince others of the necessity of environmental reform.

The purpose of ecological identity work is to provide the language and context that connect a person's life choices with his or her ecological worldview, serving as a guide that coordinates meaning, a transition to a new way of seeing oneself in the world. This work is important because it provides a moral anchor, lodged in reflective learning, a trail map for the difficult decisions that may lie ahead, a way to reiterate what's important, and a means for interpreting the experience of nature. As an educator, I have the extraordinary opportunity to build a community of learners who travel this path together.

There are some important strategic considerations here. Not all people are willing to talk openly about their values, or the critical incidents and experiences that compose their life choices, nor are they convinced that such discussions are relevant in an educational setting. And there is no doubt that ecological identity work has therapeutic implications in the sense that people want to heal themselves through their experience of nature, which may summon up painful feelings of loss as well as expressions of joy and happiness. To encourage introspection about life experience may be to open a Pandora's box. It cannot be taken lightly. It is crucial to create safe learning spaces, avenues for personal and public dialogue that allow people a wide choice in how they express themselves. What are the educational implications for ecological identity work?

A diverse group of people will only come together when they realize that they have something in common. I find there are three types of enviromental experiences that most people are interested in exploring: childhood memories of special places, perceptions of disturbed places, and contemplation of wild places. There is sufficient common ground here to stimulate dialogue, artistic expression, and mutual interest. With just a small degree of probing and encouragement, people will respond to these dimensions of their personal experience in a nonthreatening and convivial way.

And it is always helpful to construct symbolic, artistic, and experiential learning approaches as means for collective storytelling. For after all it is the stories of environmental experience that link people together. Through these stories, people recall memories and impressions of nature, and unlock the basis of their values and commitments, perhaps revealing a new interpretation, in conjunction with the stories of their colleagues. So I have people draw maps of their childhood places. I take them on hikes to secluded hilltops, have them keep creative journals which may involve writing, drawing, or poetry. They know that these documents and experiences become a collaborative text, meant to be shared with other members of the group, the basis of a mutual inquiry and a common expression. People reveal only what they will, finding a means of expression and a representation of experience that is appropriate to their own paths. These are the voices of ecological identity.

Childhood Memories of Special Places

From time to time, I have worked with international groups of environmental activists, including people from Kenya, Tanzania, Nepal, Thailand, Bolivia, Hungary, Jamaica, and Israel. They come to New England and take monthlong courses on management and conservation issues. But that is merely a rubric for a diversity of themes, interests, and backgrounds: one woman is the environmental policy coordinator for a region of her country, another woman works with an environmental law firm, one man is an elementary school teacher, another manages a wildlife preserve. Brought in as an outside speaker at some point during the program, my task is to find a way to get the group to discuss the values that inspire their commitment to environmental work.

Here is a group of environmentalists who may be from five continents, dozens of countries, with vastly different cultural perspectives, different professional interests, mixed abilities in speaking English, many of whom are in the United States for the first time. Using their common environmental interests, I have to find a way to build community within the group. But I've come to know that most people, regardless of their culture, are comfortable talking about their ancestry or their roots. They also like to talk about the places where they live, especially when they are very far from home.

To begin the workshop, I spread several dozen maps on the table and ask the participants to select the appropriate map for their home country. Dividing them into groups of four, I suggest the following sequence: use the map to describe your ancestry, where your people came from, when they arrived at your home region, how they earned their living, what the region looked like when they first arrived, what it looked like when you were a child, and what it looks like now.

A man from Tanzania has to go back only one generation to find his roots in a hunting-gathering tribe of proud warriors. He spent his childhood in the bush. A woman from Israel describes how her father escaped the Holocaust and fled to Tel Aviv; the rest of her family perished. As an urban professional, she is trying to build a relationship to the land through her environmental advocacy. This is her way of establishing new roots. With sadness and melancholy, a man from

Nepal longingly portrays the forests where he roamed as a child, describing how they had all been cut down, and in a way this made him feel rootless.

Despite the great variety of international and cultural experience, there is a striking thematic pattern: whether the person is from an Asian tropical rain forest, an African savanna, a Latin-American city, a European valley, or a North American farm, they tell a similar story. They have fond memories of a *special childhood place*, formed through their connections to the earth via some kind of emotional experience, the basis of their bonding with the land or the neighborhood. Typically these are memories of play experiences, involving exploration, discovery, adventure, even danger, imagination, and independence. And what stands out is the quality of the landscape—full descriptions, vividly portrayed, embedded in their memories.[8]

Yet these memories are often juxtaposed with a distressing contemporary picture. In many cases, the childhood landscape has been transformed by economic development. These people also express feelings of loss, despair, and frustration as their special places are irrevocably changed: sustainable, local economies are marginalized, old cultural ways are threatened, the wilderness is retreating. These perceptions are intrinsic to their environmental concerns, linked to a broader ecological understanding of threatened habitats and endangered species, as well as to a political and economic awareness of global environmental change. And perhaps the most stunning realization of all, as you listen to this catalog of international experience, is the sense that environmental deterioration is not just a local, idiosyncratic pattern. Whether it's Nepal or New York, people relate similar stories about how their special places have changed.

Here, then, are two paths for the exploration of ecological identity: memories of the special places of childhood, and the experience of disturbed places. To explore either path is to invite a journey on the other. Inevitably, we compare what we see now to what we remember of the past. But the educational context of these memories matters a great deal. To explore memory you have to be a good archaeologist, knowing where and how to dig. The purpose of revisiting the special places of childhood is to gain awareness of the connections we make with the earth, awakening and holding those memories in our consciousness of the present. Not to nostalgically pine for a lost, innocent childhood, but to recover the qualities of wonder, the open-mindedness regarding nature, the ability to look at what lies right in front of us. The purpose of witnessing the transformation of those places is to appreciate the magnitude of environmental change, to understand and feel the impact of the changes.

Gary Snyder, in *The Practice of the Wild,* describes the importance of childhood memories of place:

> The childhood landscape is learned on foot, and a map is inscribed in the mind—trails and pathways and groves—the mean dog, the cranky old man's house, the pasture with a bull in it—going out wider and farther. All of us carry within us a picture of the terrain that was learned roughly between the ages of six and nine. (It could as easily be an urban neighborhood as some rural scene.) You can almost totally recall the place you walked, played, biked, swam. Revisualizing that place with its smells and textures, walking through it again in your imagination, has a grounding and settling effect. As a contemporary thought we might also wonder how it is for those whose childhood landscape was being ripped up by bulldozers, or whose family moving about made it all a blur. I have a friend who still gets emotional when he recalls how the avocado orchards of his southern California youth landscape were transformed into hillside after hillside of suburbs.[9]

From the perspective of human development, the period of middle childhood (the ages of 9 to 12 years, perhaps somewhat later than Snyder indicates) is a time of *place-making* in which children expand their sense of self. Their perceptions of the immediate environment undergo a remarkable transformation. This may occur, as Snyder suggests, through the expanding exploration of the home territory, or through the actual creation of distinct places within that territory: dens, forts, and miniature houses—using the materials that are at hand. A child realizes, during this stage, that he or she has a unique perception of the world, one that's different from that of his or her parents, siblings, and friends. This is a time of great creativity, involving the first explorations of independence. And some theorists (including Joseph Chilton Pearce, Roger Hart, Edith Cobb, and Paul Shephard) maintain that this is a time that children establish their connections to the earth, forming an earth matrix, a terrain symbiosis, which is crucial to their personal identity.[10]

Educator David Sobel describes how for adults, special places from childhood serve as "touchstone memories," or places to which they continually return in their mind's eye. In his book, *Children's Special Places,* he explores the place-making function of middle childhood.

The roots of the adult notion of a sense of place are established during middle childhood. Rachel Carson's sense of wonder of early childhood gets transmuted in middle childhood to a sense of exploration. Children leave the security of home behind and set out, like Alice in Wonderland and Columbus and Robinson Crusoe to discover the new world. In the process, children create new homes, homes away from home. These homes become the new safe place, a small world that they create from the raw materials of the natural world and their flexible imaginations. This new home in the wilds and the journeys of discovering are the basis for bonding with the natural world. As we bonded with our parents in the early years, we bond with Mother Earth in middle childhood.[11]

It is no wonder that so many environmentalists find such depth and richness in exploring these years. Nature writing is replete with reminiscences of childhood: environmental educators carefully watch children in order to understand how they form a relationship with the natural world.[12] This is a fertile ground for the exploration of ecological identity, a way to open a window onto memory.

There are many ways to allow these memories to unfold. You can actually revisit the place where you grew up. Or you can collect old maps, photographs, documents, items of clothing, toys, and so forth that are representative of those years. Perhaps the most engaging approach, especially in a classroom setting, is to have people draw maps of their special childhood places. Some people compile specific, detailed drawings of their rooms, their neighborhoods, or particular trees or streams. Other maps convey a more abstract and impressionistic series of memories. The artistic process of drawing a map allows adults to overcome their tendency to intellectualize and categorize the experience of nature. Yet they can interpret their maps with the benefit of their subsequent experience, providing a retrospective means for understanding some of their earliest impressions of nature.

For some people, this type of reflection provides the grounding and settling impressions that Gary Snyder referred to. After drawing maps of his childhood, one of my students, Geordie, a regional planner, expressed how his awareness of place is a pillar of his life.

I am a native Vermonter. I was born here. My whole life has been in this place. Except for college, all the places I have in lived fit in a circle within a 2-mile radius. And now, as I turn 36, I live in the house I grew up in. Mount Monadnock, 30 miles away, is in the center of my view, rising symmetrically to a peak above the surrounding, rolling landscape. Throughout my life, the mountain has been there, steadfast yet full of life. Some days it is dim and indistinct, a light blue-gray mass, barely discernible at the far edge of the world. On other days the mountain is only a memory, lost in the haze. Then, as the days and seasons pass, it becomes a presence so clear and sharp it seems to be in my living room. I can measure the air from how Monadnock appears. I can see weather fronts move away from me as they cross the intervening miles. When I was 15 my sister and I agreed how unbelievably fortunate we were to live here, out of town on a dead-end road, with fields, trees, streams, and an incredible view. It is my place. I know it. I feel and love it.

Perceptions of Disturbed Places

Not everyone can easily return to those place-making years. For some, these memories are indistinct or painful. Perhaps they experienced a great deal of personal turmoil because their family was breaking apart. Or perhaps they lived in places that they would rather not remember. In discussing childhood memories of place, two of my students realized that they were in the same place at the same time, under different, but equally painful circumstances. Sabine grew up in Beirut, witnessing the horrors of a chaotic war. John was an American marine stationed in Lebanon, a so-called peacekeeper who was shocked and dismayed as he witnessed the human suffering and the destruction of the environment.

Even for those who can return to pleasant memories, a dark side will inevitably emerge. The great majority of my students have experiences similar to the international group—their childhood places are in some way polluted, developed, or destroyed, and they are overcome by impressions of emptiness and loss. Their special places may be gone, but the memories are intact. These memories serve as an idealized vision of what it feels like to bond with a place. When these feelings are attached to an ecological worldview, they provide the inspiration to recreate the same feelings in their contemporary life, an attempt to bond with the place they currently live in, and to discover adult ways of doing so. (I provide an educational perspective on this process in the sense-of-place map discussion in chapter 6.)

The stark reality is that the places they currently live in may also fall prey to development or pollution: the wetland at the end of the road, the tree across the street, the toxins in the water supply, the ozone hole in the atmosphere. Adults relate to these threats from both an emotional and cognitive perspective, integrated through the process of identification. Intrinsic to an ecological worldview is the ability to see an ecosystem as a part of oneself. This knowledge is gained both through an understanding of scientific ecology and the ability to observe and internalize the interconnections and interdependence of all living things.

Phyllis Windle's intriguing essay "The Ecology of Grief" describes how ecologists often develop a love for the organisms and places which they study. She no longer doubts her attachment to specific species, or feels alone in her grief for their loss. Ecology is a science of relationships, and through this science she has learned to identify with a wide circle of life forms and places. She observes that "ecologists are both blessed and cursed with seeing natural systems clearly" and they "see what is there and also what is gone."[13]

A great deal of contemporary nature writing contemplates these unsettled landscapes of ecological and emotional change, or what was described in an issue of the magazine *Orion* as the ecology of love and loss. To read Rachel Carson is to compare *A Sense of Wonder* with *Silent Spring*. To read Terry Tempest Williams, Stephanie Kaza, Barry Lopez, or any of dozens of contemporary writers is to explore the curious juxtaposition of wonder and doom, and to contemplate the consequences for personal identity.[14] (Chapter 5 discusses the implications of wonder and doom as it affects the environmental practitioner.)

It is important to emphasize that ecological identity work does not reflect only a sense of personal loss. One cannot separate memories and impressions of earth, community, and family: these all make up a sense of place. Robert Gottlieb, in *Forcing the Spring*, shows how the idea of place is "a dominant and powerful metaphor within the antitoxics movement, particularly for its women participants."[15] In effect, the presence of toxins and hazardous wastes in a community threatens the fabric of a community; people may lose their homes, the loss constituting a violation of place. The loss of place becomes a global issue when we consider the desperate plight of refugees, and the threats to indigenous cultures—what amounts to the breakup of the global family.

Love and loss. Wonder and doom. These are some of the feelings that emerge through ecological identity work. The educator must find a way to balance these impressions, avoiding the temptation to dwell for too long in any emotional space. For ecological identity work, the purpose of a learning community is to provide people with a forum in which it is safe to discuss these conflicting feelings, but to do so within a broader context as a way to internalize the meaning of environmentalism, provide guidance for professional and personal choices, create cohesion and solidarity within the group, legitimate the feelings themselves, and acknowledge that these discussions belong in environmental education.

It is easy to slide into depression and despair. After all, most environmentalists are motivated in part by their perception of ecological decline. To avoid this tendency, it is crucial to explore a third dimension of ecological identity work: the contemplation of wild places, for this provides the joy, wonder, inspiration, and happiness that also go to make up an ecological worldview.

The Contemplation of Wild Places

Nothing frustrates me more than discussing how to contemplate wild places while sitting indoors in a classroom. About a month into the fall semester, during the full splendor of the New England fall, I take my classes to a favorite place, Gap Mountain in southwest New Hampshire. It's a short hike—about a mile through a typical northern forest with abandoned farmlands, stone walls, early successional growth, a few big white oaks and white pines—and finally a blueberry- and boulder-covered summit. We hike silently, stopping at a few places along the way for some meditative breaks, and then at the top I give the students some time for themselves, to reflect on the place, to wander around the summit, and enjoy the view.

The group convenes in a circle. I briefly review the natural and cultural history of Gap Mountain (really a protruding, bumpy hill). The heavily forested surrounding landscape was at one time exclusively agricultural. Most of the trees are less than a hundred years old. Hence, in some respects, the landscape looks "wilder" than it might have in the recent past. Just to the east is Mount Monadnock. Even though it is 5 miles away, it appears so close that you think you could get there in one flying leap from the largest boulder. Amazingly, because of its proximity to large metropolitan areas, it is among the most-climbed mountains in the world. But from where we sit, we can't see anyone on the summit.

Further to the east is the New England coastal plain, and the densely populated areas of Boston, Nashua, and Manchester, although from our altitude and distance it all appears as a rolling plain of delicate yellows, reds, and greens. There is a low ridge of medium-sized hills that forms a boundary with the coastal plain. In the other directions we gaze at the hill country of northern New England—the White Mountains of New Hampshire, the Green Mountains of Vermont, the Berkshires of western Massachusetts. Seeing so much open space, seemingly unsettled, feels reassuring, as if we are immune to encroaching development. Yet Gap Mountain is just a hill, and we are close enough to town that we can hear the sounds of settlement: barking dogs, a factory whistle, and a reverberating chain saw. Nevertheless, by any objective contemporary standard, this is a wild place, and we can all appreciate it for being that.

I introduce what I call a *sense-of-place meditation*, guiding the students through a series of observations that allow them to focus on their senses in relation to the landscape. Feeling the air as it moves through our bodies, we contemplate the prevailing weather system. Listening to the sounds of the insects and birds, we become acquainted with the animal species. There are many variations on this theme. The point is simply to cultivate an awareness of ourselves in this wild place, to slow down for awhile and cherish the surroundings. Even those who are uncomfortable with the idea of meditation come to appreciate the experience.

The group seems to treasure this time together, bemoaning that their busy lives prevent them from contemplating the wild, and how ironic it is that the environmental profession has led them to indoor jobs, or that their thoughts about nature are so often trapped in an intellectual realm. Worried that too much discussion will put us back in our heads and away from the place, I often suggest that people spend the next week thinking about how wild places influence their lives and that they write or sketch their impressions in an *ecological identity journal*. After a few more minutes of impressions and comments, I ask the group to be silent, and we slowly make our way back to the trailhead, nourished enough by the experience that the rest of the day will look just a little bit different.

After 15 years of reading these journals, what I have found is that for many environmentalists, the direct experience of wild places has a transformational quality. Most of my students can distinguish an event, a time in their lives, or a critical series of incidents in which different strands of their lives seemed to converge, helping them carve a personal vision. Frequently, these events encompass the contemplation of the wild, or what they perceive as being "immersed in nature." By writing about these experiences and talking about them in a supportive setting, they are reiterated and validated, serving as reminders as to what's important.

Sabine was a philosophy major in college. She was living in New York City, lacking a clear sense of personal direction, confused about her career, knowing that something important was missing from her life. As she wandered home one afternoon, through the streets of lower Manhattan, she realized that she could only unlock her confusion by reconnecting with the wild. For Sabine, this couldn't happen in New York City.

> Five years of immersion in the classical texts of Western philosophy and religion left me walking on stilts, removed from the teachings of tree, rock, and soil. Living in New York for 7 years made it even harder to see past the grit of concrete and asphalt. Barry Lopez, before giving a reading at the Ninety-Second Street Y, stood at the podium and described how a great wind surged through the streets, lifting a piece of trash in swirling eddies toward the night sky. His voice and manner were infused by a sense of wakefulness and shook me from my sleep. I had gotten so numb from the abuses of living in the city that I was no longer touched by wild things. I saw my first owl in Central Park, went birding at Jamaica Bay, stood on the roof of my apartment during thunderstorms, and spent endless hours at the Museum of Natural History, staring disconsolately at wildlife exhibits, but it was not enough. I wanted to be in the thick of it. I wanted to go out west, I wanted a canopy of stars for cover, I wanted to walk down a dry wash, to grope for tracks in the dirt, to smell creosote after rainfall. I wanted to learn everything I had forgotten.

This was the event that convinced her not only to pursue an environmental career but to commit herself to a way of life that reflected her emerging ecological worldview.

Eric has spent the last 5 years as a seasonal environmental educator. He aspires to create and manage his own nature center, in a community where environmental education opportunities are not available. He grew up expecting that material affluence would make him happy, but in fact, material desires are becoming less and less relevant.

> As a child, I believed that when I grew up I would be able to have anything that I wanted. I would finish school, get a good job, and make money. I spent my childhood being enticed by the goods and services being offered through the media. Someday I would be able to afford them . . . something to look forward to. Instead I have grown up realizing that there is not time for me to have these things. These things are not realistic. I need to focus on less. It has become important for me to convince others of this too. But it is not and will not be easy to do. Bruno [a Swiss national living among the Penan in Malaysia] has given up all this. His life has become richer than I could have ever imagined in my youth. Bruno says he doesn't eat dollars. "To see something beautiful, like a mushroom, a pebble in the water, an orange salamander-these are my wages." A beefy check (although I have yet to experience this!) cannot make me feel nearly as good as I do when I sight an osprey circling over a lake or a spider's web sparkling with dew; nothing makes me feel as whole.

When Eric evokes the appeal of material simplicity, and the great pleasure he takes in observing nature, he identifies with an important aspect of the environmental tradition: Henry David Thoreau, Annie Dillard, and so many others write about their attempts to simplify their lives as a means of attaining greater observational awareness and deeper philosophical insights. Eric intimates that he must unravel the expectations of his suburban youth. The contemplation of the wild is a means to this, allowing him to reconstruct his values and grapple with his identity.

Susie, an environmental educator, describes a moment when she fully identifies with nature, perceiving herself projected onto a broad ecological and evolutionary stage:

A tropical world resides within me
Riding around on the floating shift
Of continents, cruising on the rock elastics.
I am waiting for the next collision,
Ready to exchange, mammal for marsupial.
Meanwhile, I think
All the cells in my body are replaced every two years—
Lifetime guarantee, no money back for services rendered.
Yippee, I awake feeling life,
Oh I am a different Susie.

This pure expression of happiness is a celebration of life, emerging from a profound realization. Susie experiences her connectedness to life, breaking through the narrow boundaries of a restricted ego, identifying her body with the earth, her mind with the ecosystem, her spirit with evolution. This is nothing less than the manifestation of an ecological identity, in which the object of her identification is her experience of nature.

Ecological identity work does not only occur through grand epiphanies and dramatic incidents. It is a lens through which the experiences of everyday life take on new meaning. When Priscilla returned from Gap Mountain, she noticed all the distractions in her everyday life: the mindless chatter of self-important conversations, the ubiquitous hum of advanced technology. In rare moments of silence, she has glimpses of the wild. How might she reorient her everyday life to procure more silence?

What joy, even luxury it is to walk in silence these days. So often on a trail I find myself behind people discussing business deals, work problems, and related issues in loud harsh voices, seemingly oblivious and immune to surrounding nature. How irritating it is as one tries to lag behind only to hear more loud voices coming up from the rear. I may as well be in the financial district. How sad that they miss the experience of nature that I'm sure they intended. But the mechanistic universe's hold is strong. You can hear it humming or ticktocking wherever you go, in the drone of automobile engines, the roar of jet aircraft, and the groan of the chain saw. Back home, writing in my journal, the refrigerator sings with me, the VCR whispers as it records programs on PBS that I don't have time to watch and I know the computer hums on my husband's desk in his apartment near Boston.

These are the collaborative voices of ecological identity in the sense that a variety of personal experiences constitute a collective environmental vision. While building a vibrant learning community, the educator weaves this vision so that the whole group gains greater insight regarding their environmental values, professional commitments, and social responsibility. The three avenues portrayed here—childhood memories, disturbed places, and wild places—are paths of exploration, educational possibilities for the collective expression of ecological identity—fields of wildflowers, if you will, comprising a much broader landscape. Ecological identity is a conceptual means for exploring the patterns of personal and professional growth, as linked to the broader context of the environmental tradition, everyday life, global issues, and environmental politics. As the book proceeds, I describe these links in much greater detail, exploring educational paths for these domains. But first, it is necessary to place this concept, ecological identity, in a broader context, as it is merely one of many interesting approaches that are working toward similar ends.

Ecological Identity Amplified

Ecological consciousness, ecosophy, the ecological self, the ecological unconscious: these are just a few of the metaphorical terms that have been used to formulate an epistemology of mind and ecosystem. They are epistemological in the sense that they offer an approach to knowledge based on an understanding of ecological concepts, not only as they are derived from scientific ecology but also from vernacular cultures and ancient philosophies. These formulations involve both construction and critique. In other words, they offer a new synthesis of knowledge, based on a comprehensive reappraisal of various normative views of the world. It is beyond the scope of this book to fully describe the challenging philosophical questions these concepts imply. The purpose of this short survey is to consider how these ideas support the notion of ecological identity, and to reflect on the educational implications.

In 1977, Herbert L. Leff, an environmental psychologist, wrote a landmark text called *Experience, Environment, and Human Potentials.* In this ambitious work, Leff asks what types of conditions and environments are most likely to support the full realization of human potential, and he does this assuming that what he calls "ecological consciousness" is intrinsic to this process. Also, Leff proposes a series of theoretical and practical cognitive suggestions to enhance creativity, self-awareness, and personal change, as these qualities reflect environmental values.

Leff posits the idea of ecological consciousness as a "specific ideal cognitive, valuational, and motivational orientation toward the world . . . more a goal to strive for than a particular pattern of psychological functioning that we can readily investigate."[16] He suggests that ecological consciousness is comprised of four chief components: ecological systems thinking, a high ability to enjoy and appreciate things in themselves, an ecocentric value system, and a synergistic orientation in interactions with one's social and physical environment. Leff elaborates these components in full detail, listing the specific behavioral implications.

Perhaps most relevant for ecological identity is his discussion of ecological systems thinking, which involves two qualities: a high level of ecological understanding and awareness and the sense of self as part of a larger system. He suggests that the ability to see oneself as an integral part of the biosphere is difficult to internalize, nevertheless "it seems to offer the key to an active awareness and appreciation of our essential unity with each other and the rest of nature."[17] Further, Leff explains how ecological understanding has both an intuitive and cognitive dimension.

Besides an intuitive appreciation of one's own underlying "identity" (or at least integral connection) with the biosphere, ecological systems thinking would involve both an understanding of ecological processes and a continuing awareness of how those processes operate in one's own life and surroundings. Thus, you could know a lot about ecological processes and still fail to use cognitive sets that would actively relate this knowledge to your ongoing environmental experiences. Conversely, you might try to attend to ecological processes in your environment, but such a cognitive set might well prove impossible or misleading if you did not understand ecological principles.[18]

For example, consider how a person begins to understand the intricate relationships in the hydrological cycle. There is the cognitive dimension, as one might learn it from an environmental science text. You learn that water goes through a series of complex transformations, from clouds to rain, through the ground, in an aquifer, through a watershed, down a river, into an ocean, and back to the atmosphere. It evaporates and transpires; becomes solid, liquid, or gaseous, combining with other minerals, while passing through animals and plants. This knowledge is fundamental to environmental science, and similar to other biogeochemical cycles (carbon, nitrogen, sulfur, etc.) depicts the complex and intricate interrelationships between organisms and their environment.

Another path to this same knowledge lies in the direct experience of nature. By watching the sky, observing the land, following the watercourses, and noticing how living things use and depend on water, one can derive a very sophisticated intuitive understanding of the relationship between water and life. Through sensory awareness and the contemplation of nature, one can appreciate this complexity and intricacy, without naming it the hydrological cycle, or referring to the various chemical, physical, and biological laws.

An understanding of biogeochemical cycles conveys interconnectedness and interdependence, critical aspects of an ecological worldview. My impression is that people may arrive at this perspective from two different, but connected paths. For some, it is a cognitive understanding of scientific ecology that leads them to this view. They use this knowledge as a means to extrapolate principles for living and as a way to understand their place in the world. Yet some people arrive at notions such as interconnectedness and interdependence from a purely experiential perspective. They have had experiences in nature or connections to the earth that have allowed them to understand ecological relationships from a more intuitive approach. Their knowledge of scientific ecology may be anecdotal and metaphorical. Of course these different approaches are not polarities; they merely reflect two different ways of knowing. What is crucial is how these approaches inform each other.

And there is the subsequent question: how do people internalize their ecological understanding so as to expand their sense of self? Arne

Naess, in *Ecology, Community, and Lifestyle,* elaborates a philosophy that "leads from the immediate self into the vast world of nature" with the intention of allowing people to "find ways to develop and articulate basic, common intuitions of the absolute values of nature which resonate with their own backgrounds and approaches."[19] This comprises *ecosophy,* " a philosophical world-view or system inspired by the conditions of life in the ecosphere." For Naess, self-realization occurs as a process of identification with nature. He describes the ego as a fragment of a larger whole, and through our identification with greater wholes, we see that we are more than just our egos, but aspects of a more inclusive process, the biosphere itself. As we realize this, we experience new dimensions of satisfaction and explore new levels of meaning. "The ecosophical outlook is developed through an identification so deep that one's own self is no longer adequately delimited by the personal ego or the organism. One experiences oneself to be a genuine part of all life."[20]

What are the perceptual consequences of this more inclusive, more ecological sense of self? Joanna Macy, in *World as Lover, World as Self,* believes that when people make these broader connections, they begin to perceive damage to the biosphere as coextensive to themselves. As people identify with the planet, they realize that the world can be perceived as their body. The trees in the Amazon rain forest represent the lungs of the planet and their demise is a blow to the personal and global respiratory system. Macy describes this realization as the *greening of the self.* She argues that conventional notions of self are being peeled off as people experience the extent of environmental pollution, replaced by "wider constructs of identity and self-interest—by what you might call the ecological self or the eco-self, coextensive with other beings and the life of our planet."[21]

Both Naess and Macy are cautious and skeptical about the moral exhortation and intellectual reductionism that may accompany celebrations of the unity of life. The purpose of broadening the concept of self is not to lose individuality, but rather to widen the possibilities for ecological awareness. Both emphasize the metaphorical implications of identification, and use revised notions of self as a means to facilitate social change and environmental reform. Naess describes ecosophy as a constellation of approaches, designed to expand the parameters of personal choice and motivate people to take environmental action. Macy describes the self as a metaphor which we can creatively expand as we choose. Identification and differentiation are synonymous. That is, as people conceive of wider circles of identification, they also realize the ways in which they are *different* from other life forms, how their human experiences constitute a unique path. From this perspective, the wider our circles of identification become, the more we extend the possibilities of self and the more liberated we become in how we perceive ourselves in relation to nature.

Macy's work involves a series of comprehensive guided meditations, designed to facilitate the emergence of the ecological self. She addresses the various dimensions of stress and anxiety that accompany perceptions of ecosystem decline. Inevitably this material has therapeutic implications, as it encourages significant introspection and critical reflection. Her position is that only a resurrected ecological self can repair the wounds that humans have inflicted on nature. Thus the individual must undergo a healing process in order to fully understand his or her relationship to the earth.

Hence the emergence of ecopsychology, a therapeutic orientation that emphasizes a simultaneous healing process: ecological restoration accompanies personal reconnection. Theodore Roszak, in *The Voice of the Earth,* develops the metaphor of an ecological unconscious, suggesting that "the collective unconscious, at its deepest level, shelters the compacted ecological intelligence of our species . . . it is this id with which the ego must unite if we are to become a sane species capable of greater evolutionary adventures." [23] Roszak proposes the preliminary working principles of ecopsychology. Although his guidelines are highly generalized and relatively nonspecific, he suggests that ecopsychology focus on the life of the child, utilize the traditional healing techniques of primary people, emphasize the experience of wilderness, demystify sexual stereotypes, and "awaken the inherent sense of environmental reciprocity that lies within the ecological unconscious." For Roszak, the purpose of ecopsychology is to create a new domain of study which liberates people to discover the various ways that they can connect with the earth.

Ecological Identity Applied

I view these ideas as experimental learning tools—convergent interpretations that allow people to think about the broader context of ecological identity work. Conceptual tools serve as bridges that connect diverse experiences. Theories provide temporary meaning, trailmarkers along the path of life experience. Within each of these metaphors there is room for a great deal of intellectual exploration and experiential learning.

Ecological identity work yields a rich substrate, prompting critical reflection and deep introspection, a kind of personal awakening which allows people to bring their perceptions of nature to the forefront of awareness and to orient their actions based on their ecological worldview. I observe two simultaneous unfolding processes. Widening the circles of identification proceeds as a form of exfoliation, a peeling away of layers, a breaking of perceptual boundaries, allowing for more expansive circles of awareness. Opening the windows of memory prompts looking deep within, as one might peel the layers of an onion, the removal of each layer bringing you closer to the core. These are dynamic, connected, concentric circles, with each layer representing another aspect of ecological identity.

Moving through these concentric circles might cause a great deal of personal change. As you probe the layers within, you might realize that your experiences in nature are the source of profound wisdom and personal happiness. What are the implications for your everyday life—your job, your personal relationships, your professional goals? Perhaps it becomes necessary to rethink some aspects of your life which you previously took for granted.

As you widen the circles of identification and realize that your sense of self includes the regional watershed, and you internalize threats to water quality as you never have before, do you become more involved in politics or talk about your concerns with family and friends? Perhaps it becomes necessary to reorient your time commitments and personal affairs so that you can explore these feelings and translate them into community action.

Ecological identity work also involves concentric circles of tension and satisfaction. As you realize the importance of being present in nature, you might also observe how often you are absent from nature, ever aware of the distractions that prevent you from paying attention. As you discover how much you require solitude and retreat to contemplate nature, you might realize that you have a responsibility to get more involved in time-consuming political action. And as you more clearly formulate a strong environmental point of view, you might observe how your behavior and actions challenge other people.

These issues inevitably emerge in my classes because they represent what people want to talk about—the practical day-to-day implications of ecological identity work. Whether you are an environmental professional or a concerned citizen, the domain of environmental issues is widening. The theater of practice is ubiquitous: the halls of public policy, consumer habits, the school curriculum, the business world. The possibilities of ecological identity reverberate through the commonplace settings of the moment. People want to talk about how an ecological worldview can guide them through the situations of everyday life. Without these applications, ecological identity work is just empty talk and hollow reflection.

In forthcoming chapters, I re-create many of these conversations, trace them through the circuitous path of ecological identity, reveal the collaborative insights of professional environmentalists and concerned citizens. In the next chapter, I look more carefully at the environmental movement, analyze some of its prominent historical figures, and explore the broad spectrum of environmental thought. That is the next phase of ecological identity work: the ability to understand how one's personal values are reflected in the historical and political emergence of contemporary environmentalism.

2 Trees of Environmentalism (Ecological Identity Evolving)

Environmental Trees as Portraits of Ecological Identity

Picture environmentalism as a tree. Note the complex network of roots, trunk, leaves, and branches. The roots represent the seminal ideas—the philosophical legacy and moral foundations, including the great teachers and role models, the classical books and essays. The branches reflect the evolution of those ideas, their multiple variations. They form a network of interpretations, in some cases revealing heated controversies, in others the convergence of separate paths. Then there are the people, organizations, and perspectives that make up contemporary environmentalism. These are the leaves of the tree, spreading out in a broad canopy, signifying the diversity of the movement. In the trunk, where the sap flows through the system, you perceive the dynamic interchange of ideas and actions, linking the present to the past, connecting the self to the world (figure 2.1). This is where your ecological identity unfolds.

When I teach courses about environmentalism, I use this metaphor as the basis for an extended project and a framework for integrating knowledge and personal values. After a lengthy period of reading, thinking, and discussing the origins of contemporary environmentalism, I ask my students to create an artistic representation of a tree, comprising a personal view of the broad spectrum of environmental thought, including any people, ideas, events, or texts that have significantly influenced their approach to environmentalism. I encourage them to use creative expression in developing the metaphors and symbols for their trees. In effect, anything goes. And over the years I have seen an extraordinary range of presentations: annotated drawings, charts, illustrated essays, sculptures, paintings, photographs, and mobiles.

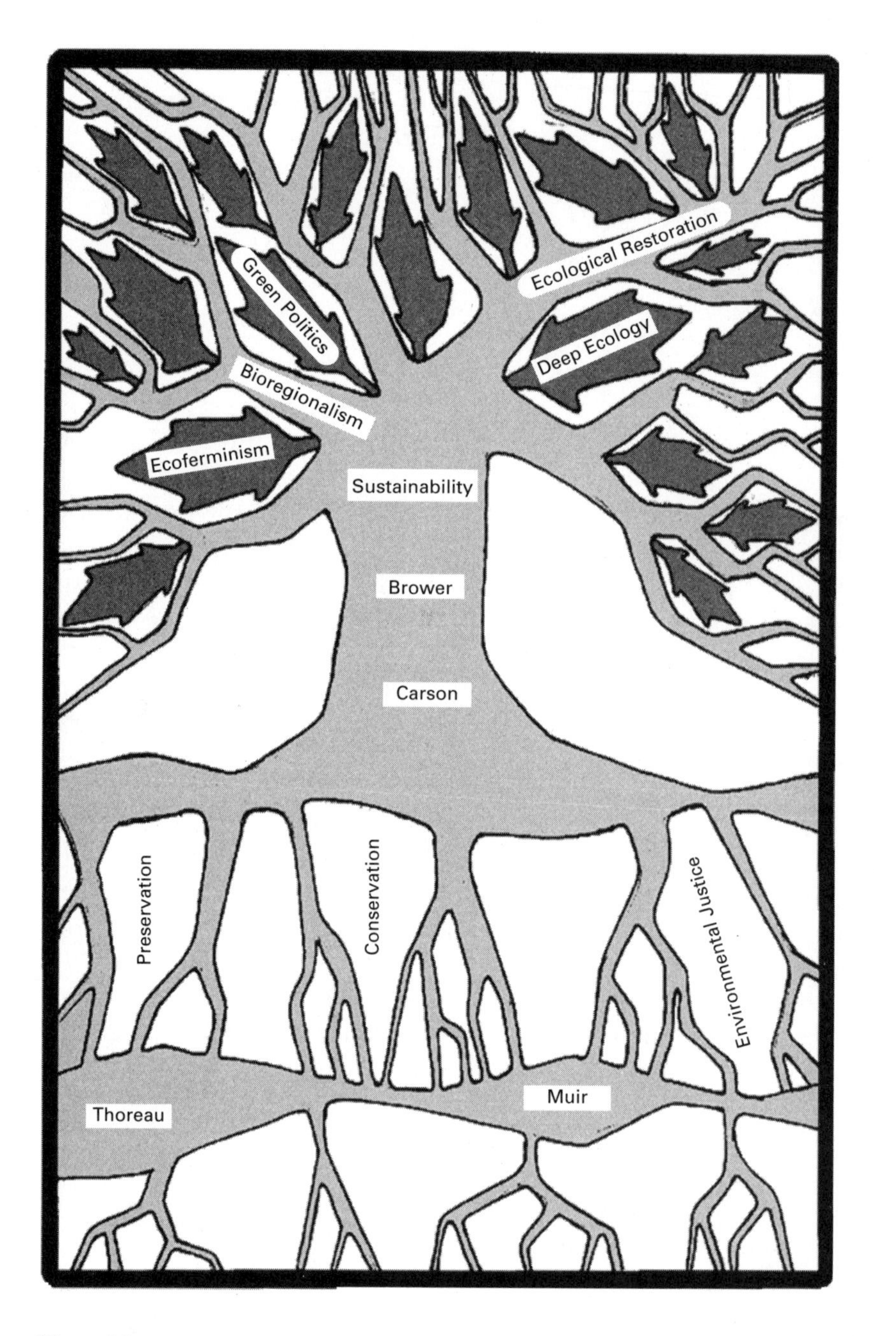

Figure 2.1
A tree of environmentalism

This project involves three interpretive challenges. First, the student must figure out how to synthesize this information. Which groups and ideas go where? Although several historians have developed useful taxonomies of environmentalism (which I briefly describe later), it is not easy to place this material in an organized framework. Second, I encourage the student to place himself or herself on the tree. Where do you belong in this picture? Many different perspectives are depicted in these projects, and a person makes a strong personal statement through his or her identification with a particular approach. Third, the tree reflects how a person conceives ecological identity. In what ways has your understanding of environmentalism influenced your sense of self? The artistic approach allows people to express themselves in a nonlinear fashion, moving beyond the written word, integrating their knowledge of environmentalism with their deepest beliefs, feelings, and motivations.

When the assignment is due, students hang their trees on the wall, along with the creations of their colleagues. The walls of the class become a picture gallery. These are unique artistic expressions, each providing a different glimpse of, a personal statement about, the American environmental tradition. I have now seen hundreds of these trees, yet in each new class I am delighted by the depth and variety of expression. Each tree reveals an interesting angle and perspective, an interpretation of environmentalism that I have not quite seen before.

After the class has surveyed all the trees, a lively discussion ensues. Many students are struck by the diversity of intellectual expression and the magnitude of their differences. One person asks another why "deep ecology" is so prevalent in his tree, wondering whether such a radical perspective can be at all practical. Another person explains the importance of women's issues and ecofeminism in her life, explaining how these ideas influenced her personal decisions and inspired her to pursue an environmental career. Another woman created a stunning tree out of hundreds of pieces of colored construction paper. Her leaves portrayed dozens of ideas and organizations. She couldn't differentiate between the sections of the tree, feeling that she belonged in all of them simultaneously.

There is strength and controversy within this diversity. Although the collective tapestry yields a powerful sense of affiliation— the environmental tradition appears dynamic and strong—it also reveals the important philosophical differences. There is a long leap from a conservationist like Theodore Roosevelt to an ecofeminist like Charlene

Spretnak. Yet clearly, both emerge from the American environmental tradition. There is a vast political difference between the legislative work of an EPA bureaucrat and the civil disobedience of an EarthFirst!er.

The metaphor of the tree encompasses this diversity. Leaves are photosynthetic, creating food for the tree, providing a dynamic flow of nutrients and ideas. The leaves on each tree have different shapes and sizes, denoting the breadth of people and organizations. Leaves fall to the ground and decompose, are eventually reabsorbed in the roots, changing how the past is understood, providing new perspectives on what's important. The new is at the top and the old is at the bottom, but what was once new becomes old and is enfolded in the interpretive tradition. The trunk is where this process becomes sorted out, where the flow and mixture of ideas is most dynamic, where the contrast between stability and change is most pronounced. Perhaps this is the way thoughts travel through the self—constantly changing, influencing the formation of ecological identity in relationship to the influx of knowledge and ideas. As the class observes these trees, they witness a multiplicity of interpretations, noting the interlocking themes, how they diverge and converge, where the crosscurrents are most powerful. Each tree is a manifestation of ecological identity that allows the students to see themselves and one another reflected in the *emerging interpretive tradition* of American environmentalism.

These terms are important enough that they should be more fully defined. *Tradition* means that environmentalism moves from one generation to the next, both as a series of beliefs and through the habits of community life. Knowledge is passed on from one person to the next, either through books and ideas, or through the ecological practices of everyday life. *Interpretive* signifies that people look to environmentalism as a way to formulate an ecological identity, to learn about appropriate ways of living in nature, to make sense of the world around them, and to construct a moral point of view. *Emerging* describes how environmentalism is still in its formative stages. It is a movement that constantly takes new shapes as it generates new ideas. Environmentalism and its practitioners evolve in this intricate dialectic, using ecology as both science and metaphor, lending coherence to human-nature relations.

In these trees, the theme of ecological identity is most pronounced. Through their exposure to various expressions of ecological identity, the students amplify, challenge, nourish, and legitimate their voices—

they proclaim their affiliations. In this chapter, I explore how ecological identity unfolds in diverse corners of environmentalism. I review the lives of Henry David Thoreau, John Muir, and Rachel Carson within the framework of ecological identity, how their experiences have provided guidance and inspiration for generations of environmentalists. Then I survey the preservation/conservation debate, in an attempt to reveal dramatically different perspectives about nature, leading to diverse political and professional approaches. Finally, I mention some of the innovative, contemporary approaches to environmentalism that lend great insight into ecological identity. Throughout the chapter there is a common theme: *in formulating an ecological identity, it is crucial to understand where you fit in the broad spectrum of environmental thought*. This awareness helps coordinate ecological identity with professional direction and personal choices.

Roots: Thoreau, Muir, and Carson as Environmental Archetypes

Role Models for Ecological Identity

Walden Pond, the High Sierras, a Maine seacoast cottage. At these sites, three remarkable figures of environmentalism formulated their ecological identity. For Henry David Thoreau, a man of the early nineteenth century, Walden Pond served as an enormous reflective mirror: living simply, contemplating nature, learning natural history, attempting to expand his sense of self in relationship to nature. Approximately 50 years later, in the late 1800s, John Muir ascended the High Sierras, fully extending himself by exploring the mountains, searching for the wildness in nature and finding the wild within. On the most exposed promontories, he experienced the fear and exhilaration that inspired wider circles of identification with nature. In the middle of the twentieth century, Rachel Carson, through her passion and zeal for literature and science, merged a cognitive and intuitive understanding of ecological systems, as she was fully immersed in the microhabitats of the edge of the sea. Her feelings of wonder and love became starkly contrasted with impressions of pain and loss, as she warned Americans about the devastating threats to nature.

These figures, encompassing 150 years of historical experience, have had a profound impact on American environmentalism. The quality of their life experience was so rich, their challenges so profound, and their struggles so deep, that they became figures that now appear larger than life. Thoreau, Muir, and Carson are archetypes of American environmentalism. Their life experiences serve as original patterns, historically reiterated through countless individuals. By virtue of their life stories, many environmentalists can validate their own seemingly idiosyncratic experience. Although Thoreau, Muir, and Carson are products of unique historical and cultural settings, their challenges and insights transcend their times because they grappled so intensely with the formulation of ecological identity. Indeed, we can read their lives as attempts to expand their senses of self, to widen their circles of identification, by becoming fully immersed in nature. And in each case, they dealt with the tensions, contradictions, and frustrations of applying that perspective to personal growth and public choices. For this reason, among others, their books and essays have become holy texts of environmentalism, subject to diverse interpretations, timeless in content, and wide-ranging in appeal.

Thoreau, Muir, and Carson (whom I will refer to at times as the environmental archetypes) attempted to live according to what they perceived as the moral principles of nature, placing themselves against the mainstream, often without anchors, in contrast to the habits and mores of their contemporaries. They took difficult risks, plunged into the unknown, demonstrated extraordinary courage and vision. A role model is someone whose actions, commitments, life choices, and moral integrity serve to guide others through their own life paths. This section describes why their visions are intrinsic to environmentalism.

In chapter 1, I offered several educational paths to ecological identity: childhood memories of place, perceptions of disturbed places, and the contemplation of wild places. In this section, I elaborate on these approaches. There are several salient, recurring themes in the history of American environmentalism that inspire people to reflect on their ecological identity. The environmental archetypes provide guidance in this regard. First, is what Gary Snyder refers to as the *practice of the wild*, or learning how to live using wild nature as a guide and teacher.[1] Ecological identity perennially revolves around this path: cultivating sensory awareness, locating the wild in everyday experience, using nature as a classroom for the lessons of life. Second, and closely related, is the *natural history excursion*—the wilderness journey, the natural history field trip, the study of nature in the backyard or vacant lot—these approaches serve writers, educators, and others alike as a means of observing natural history, learning about ecology, and also discovering threats to the environment. Third is the *path of citizenship*, the professional and political obligation to educate the public about the threats to nature, and the various types of involvement that result. This is a source of conflict and tension, and created a great deal of turmoil for Thoreau, Muir, and Carson, just as it does for contemporary environmentalists. Inevitably, as people widen their circles of identification, they become involved in thinking about the commons, the aspects of nature that everyone shares, and typically, this leads to participation in the public arena.

The Practice of the Wild

The environmental archetypes, in their unique ways, oriented their lives so they could spend as much time as possible in wild places—to rediscover the animal within themselves, incorporating their bodies within the earth organism. They would eat, sleep, think, and be nature.

Thus John Muir climbs tall trees in the middle of wild wind storms and pushes boulders down the sides of inaccessible cliffs. Muir searches for the rocky precipice so he can dangle, grasping the full richness of his life against the backdrop of a spectacular Sierran landscape. He describes the clarity that ensues when he faces imminent danger during one of his more treacherous mountaineering escapades:

> After gaining a point about halfway to the top, I was suddenly brought to a dead stop, with arms outspread, clinging close to the face of the rock, unable to move hand or foot either up or down. My doom appeared fixed. I must fall. There would be a moment of bewilderment, and then a lifeless rumble down the one general precipice to the glacier below.
>
> When this final danger flashed upon me, I became nerve-shaken for the first time since setting foot on the mountains, and my mind seemed to fill with a stifling smoke. But this terrible eclipse lasted only a moment, when life blazed forth again with preternatural clearness. I seemed suddenly to become possessed of a new sense. The other self, bygone experiences, Instinct, or Guardian Angel—call it what you will—came forward and assumed control. Then my trembling muscles became firm again, every rift and flaw in the rock was seen as through a microscope, and my limbs moved with a positiveness and precision with which I seemed to have nothing at all to do. Had I been borne aloft upon wings, my deliverance could not have been more complete.[2]

Muir is fully alive when he gathers and integrates all of his senses. He has integrated his mind and his body, but he has done so in an awesome and spectacular mountain landscape, his temple of challenge, his trail of grandeur. Although Muir realizes that he may have overstepped his bounds, that he is temporarily paralyzed by fear, he is awakened by this danger to attain extraordinary clarity and vision.

As Muir traverses and celebrates the mountains, as he exposes himself to the harsh clarity of blizzards and thunder, of empty skies and dazzling heights, Thoreau relishes the swamp, the "impermeable and unfathomable bog" where his "spirits infallibly rise in proportion to the outward dreariness." "When I would recreate myself, I seek the darkest wood, the thickest and most interminable and, to the citizen, most dismal swamp. I enter a swamp as a sacred place, a sanctum sanctorum. There is the strength, the marrow of Nature."[3]

We can easily imagine Thoreau sucking from that marrow; deriving nourishment from the thick, life-filled waters; bathing in the muck; diving into the bog. Thoreau explains that the health of a town is more a function of the woods and swamps that surround it than the righteousness of its people. The swamp is a mysterious landscape, undomesticated, unplowed, and unheeded as well. It fascinates Thoreau precisely because of its inaccessibility. It enables him to locate himself in nature by finding its deepest corners and its hidden domains. In the murkiness of the swamp, Thoreau's mind is most clear.

Thoreau showed that this clarity and presence could also become the staple of everyday life. Nature can be experienced in the most mundane circumstances, transforming the ordinary into the sublime. That is his quest, the desire to discard the shackles of town habits and discover the wildness that permeates the interstices of daily living. To allow that wildness to reign supreme, one must be free enough to indulge in the exquisite purity of the senses. This is how he teaches himself to see things from a fresh perspective, from a wild perspective. Thoreau found this wildness everywhere. In the ubiquity of wildness he found his salvation. In *Walden*, Thoreau lets housework become a "wild" endeavor.

> Housework was a pleasant pastime. When my floor was dirty, I rose early, and, setting all my furniture out of doors on the grass, bed and bedstead making but one budget, dashed water on the floor, and sprinkled white sand from the pond on it, and then with a broom scrubbed it clean and white; and by the time the villagers had broken their fast the morning sun had dried my house sufficiently to allow me to move in again, and my meditations were almost uninterrupted. It was pleasant to see my whole household effects out on the grass, making a little pile like a gypsy's pack, and my three-legged table, from which I did not remove the books and pen and ink, standing amid the pines and hickories. They seemed glad to get out themselves, and as if unwilling to be brought in. I was sometimes tempted to stretch an awning over them and

take my seat there. It was worth the while to see the sun shine on these things, and hear the free wind blow on them; so much more interesting most familiar objects look out of doors than in the house. A bird sits on the next bough, life-everlasting grows under the table, and blackberry vines run round its legs; pine cones, chestnut burs, and strawberry leaves are strewn about. It looked as if this was the way these forms came to be transferred to our furniture, to tables, chairs, and bedsteads—because they once stood in their midst.[4]

Thoreau experiences the wild by constructing this strange juxtaposition. He cleans his house with the sand from the pond. He animates his desk and his books by placing them among the trees. It is apparent that his table is made of wood, and it is when they are both outdoors, Thoreau and the table, that Thoreau can recognize the table for what it is, surrounded by nature, embedded in context. The purpose, function, and aesthetic of his relationship to the table becomes clear. They are all the same thing, made from the same materials, different manifestations of nature. Elsewhere, Thoreau writes about walking through diverse landscapes, tasting wild edibles, seeing the world through the eyes of a muskrat, serving as a self-appointed "inspector of snowstorms and rainstorms": his essays and journals revel in the delicious tastes, smells, sights, and sounds of the natural world. His delight is profound; he is most happy when he is most wild, when he can proudly proclaim himself an animal, because at that point he truly knows himself, he is united with his origins and thus he is present in the wild moment.

The *Walden* passage reminds us that wilderness is a state of mind. It is not necessary to ascend a dangerous peak or take a long backpacking trip to practice the wild. It is a matter of how you see the world, how you perceive the artifacts of everyday life, whether you notice the sky and hear the birds. Is it poor housekeeping to let spider webs grow in the corner of your room? Is it crude and uncivilized to eat meals with your hands? The point is not to idealize the wild or to judgmentally set the wild apart from civilization as polar opposites. Rather these experiences are reminders of the proximity of the wild, that life has an ecological and evolutionary context, that the things people use (natural resources) are transformed from the wild, and that as sure as you live you will also die and return to the earth from where you came. The wild is a pathway to contemplation, learning, and wonder.[5]

For Rachel Carson, practice of the wild was immersion in microhabitats, and her delight was exploring the behavior and environments of living creatures. Her finest writing is about her serendipitous experiences as a naturalist, when her senses became fully awake through the intensity of her observations. In these moments, she expanded her sense of self to include the interconnectedness of all life. This theme prevails particularly in *The Edge of the Sea,* which explores the interface between land and water, and in so doing reveals Carson's explorations of mind and ecosystem, of the magnificent mystery of life.

> The shore at night is a different world, in which the very darkness that hides the distractions of daylight brings into sharper focus the elemental realities. Once, exploring the night beach, I surprised a small ghost crab in the searching beam of my torch. He was lying in a pit he had dug just above the surf, as though watching the sea and waiting. The blackness of the night possessed water, air, and beach. It was the darkness of an older world, before Man. There was no sound but the all-enveloping, primeval sounds of wind blowing over water and sand, and of waves crashing on the beach. There was no other visible life—just one small crab near the sea. I have seen hundreds of ghost crabs in other settings, but suddenly I was filled with the odd sensation that for the first time I knew the creature in its own world—that I understood, as never before, the essence of its being. In that moment time was suspended; the world to which I belonged did not exist and I might have been an onlooker from outer space. The little crab alone with the sea became a symbol that stood for life itself—for the delicate, destructible, yet incredibly vital force that somehow holds its place amid the harsh realities of the inorganic world.[6]

Reading Rachel Carson resembles a slow walk on an ocean coastline. Through patient observation, you encounter diverse habitats, a variety of life forms, the mysterious interpenetration of land and sea. As you observe more closely, and become more focused, the details of your walk become increasingly pronounced. The further you explore, the more you are willing to become totally immersed in your surroundings, the greater the possibility of learning and insight. And then there is an opening, a glitter, an extraordinary moment when more than you ever thought possible is revealed. These are the unpredictable moments, when your mind is clear, and the environment is a vast reflection. But this cannot be achieved without discipline and effort. Rachel Carson revealed for millions of readers that one can practice the wild by patiently studying the extraordinary habitats of the natural world.

The Natural History Excursion

Thoreau, Muir, and Carson used excursions in their desire to locate the wild. They explored both dramatic journeys (trips to exotic places), directed study (the natural history field trip), and ordinary circumstances (nature in the backyard), seeking the wild wherever they went, showing their readers how to look and where to look, teaching them that one can experience the wild anywhere if sufficiently unencumbered and adequately prepared. For example, consider some of Thoreau's educational motivations for the Walden Pond experience:

"I went to the woods because I wished to live deliberately, to front only the essential facts of life, and see if I could not learn what it had to teach, and not, when I came to die, discover that I had not lived."[7]

Thoreau was the original environmental educator. He experimented with different perceptual approaches, traveling to the Maine woods and Cape Cod, living in a cabin at Walden Pond, using his powers of observation and interpretation, playing with consciousness, contemplating the different ways one can be in nature. Donald Worster, in *Nature's Economy*, explains how Thoreau tried to integrate "indoor science and Indian wisdom."[8] Sometimes he would be the patient naturalist, painstakingly identifying the local flora. He would also throw his books aside, searching for the total apprehension of intuitive knowing.

The metaphor of walking emerges as Thoreau's most integrated approach to wild practice. His "Walking" essay can be read as a learning tool, a guide to various methods of learning about nature. How a person walks reveals how he or she will move through life. Thoreau tells us that he must walk to live, for the process of walking is a means for observing nature and a path to self-discovery.

"I think that I cannot preserve my health and spirits, unless I spend four hours a day at least- and it is commonly more than that- sauntering through the woods and over the hills and fields, absolutely free from all worldly engagements."

He tells us how to walk. "Moreover, you must walk like a camel, which is said to be the only beast which ruminates when walking." He can walk every day in the same vicinity and still discover something new. "An absolutely new prospect is a great happiness, and I can still get this any afternoon. Two or three hours walking will carry me to as strange a country as I expect ever to see. . . . There is in fact a sort of harmony discoverable between the capabilities of the landscape within a circle of ten miles radius, or the limits of an afternoon walk, and the threescore years and ten of human life. It will never become quite familiar to you."[9]

Every walk holds the prospect of discovery, a journey to lands familiar or new, revealing new habitats and places, both around and within. It is in this way that Thoreau initiates his search for ecological identity, walking in a circle, surrounding himself with the woods, availing himself of the wild. When he walks, he feels, touches, and experiences the landscape, cognizant that he is involved in a daily practice, seeking renewal, finding balance and equilibrium.

John Muir took Thoreau's advice very seriously, also taking long walks and extensive journeys. In 1867, 29 years old, he undertook a thousand-mile walk from Kentucky to the Gulf of Mexico. When he was in his thirties, he thoroughly hiked and explored the mountains of California. At 40 he made the first of several visits to Alaska. Frederick Turner, in his biography of Muir, describes these journeys as one man's singular rediscovery of America. It was in the wilderness that Muir confronted the "inseparability of life and death and their reconciliation in nature," contemplated the prospect of plant and mineral sentience, recognized the stunning interdependence of all organisms, and became a skilled naturalist.[10] On these journeys he realized his moral obligation to become politically active in order to protect the American wilderness.

For Muir the desire to roam, explore, and discover was also the means through which he gained personal confidence. Although his motivations were complex (he suffered an abusive childhood, limited by the overbearing religious authority of his strict father; and was nearly blinded in a machine accident), he was driven to undertake dramatic journeys across diverse landscapes. These were not just a medium for reverie and contemplation. Muir tested himself severely, placing himself in the most challenging conceivable physical situations, taking great delight in his ability to resolve them. The wild was also a test of his manhood and courage.

John Muir will forever be associated with the Sierra Club, an organization that was founded on the principle of the wilderness outing. To this day, despite its enormous organizational growth and its prominent advocacy agenda, the Sierra Club promotes a bounty of wilderness trips for a diverse clientele of environmentalists, based on the belief that to protect the wilderness, it is essential for people to experience it for themselves. For that reason, the Sierra Club and many other environmental organizations emphasize outdoor experience as a way to engage people in a more personal search, hoping that out of such exposure, a new sensibility will emerge. The prospect that a person will discover nature through a wilderness experience is intrinsic to the environmental tradition and a critical foundation of environmental education. Muir's legendary journeys provide the backdrop for a century of environmental excursions. Perhaps some of these trips are now commercialized and adulterated, but the motivation remains the same: people are searching for their relationship with nature, and in their own ways hope to become enlightened in a wilderness environment. American environmental literature is replete with stories of men and women finding themselves through wilderness excursions, describing incidents critical to their personal development.[11]

Rachel Carson epitomizes the local conservation project. She provided the inspiration for thousands of concerned citizens to organize themselves, perform their own research, "marshal the experts," and to see the health of the ecosystem as intrinsic to community health. When the conservation commission performs water sampling, when the high school science class measures air quality, or when an Audubon Society leads a trip to a threatened habitat, these trips emerge from the example of her work. These excursions integrate environmental science, natural history, and community politics. One must observe natural history not only for the sheer joy of it but also to survey the ecological health of an area, and to proclaim the results to a public audience.

For most people, these excursions are important learning experiences, not just for the personal challenge and the sense of adventure but for the experience of observing natural history and understanding ecological relationships. Thoreau, Muir, and Carson were superb naturalists, keen observers of the botany, geology, meteorology, and field ecology of whatever domains they explored. Their essays were built around their observations of natural history. These were not merely metaphorical ploys, although the recursive quality of mind and nature is always apparent. Rather they had an unquenchable thirst to know about the natural world. Their essays are filled with keen ecological observations.

Thoreau anticipates modern ecological thinking when he muses about various taxonomy schemes, feeling limited by the Linnaean system and wondering whether species should be organized by habitat. When Muir and Carson observe the intricate interconnections between organisms and their environments, they are reiterating basic ecological principles. Carson was a pioneer in warning the American public about the threats of pesticides, and she discovered this through her expertise in natural history. Thoreau, Muir, and Carson simultaneously observe, categorize, interpret, and theorize. They skirt the boundary between analysis and intuition, knowing that to name is to know, but one need not always name. They kept copious field notes and compiled comprehensive field guides, but they were equally comfortable without them, and would toss them if the mood demanded.

The affinity for natural history that links Thoreau to Muir to Carson is also the glue that binds the American environmental tradition. For decades the natural history essay has been employed to express ideas about nature, education, philosophy, and politics. Much of the classic and influential environmental literature has taken this form. To this day, nature writing is one of the most popular expressions of environmentalism. Edward O. Wilson, in his stunning and highly influential book, *The Diversity of Life,* which proclaims the importance and possible demise of ecological diversity, uses his experience as a naturalist and his sensibility as a philosopher to allow his science to become lively and accessible.[12]

The Path of Citizenship

In chapter 1, I described how the formulation of ecological identity involves the perceptions of disturbed places. This is an inevitable dilemma for environmentalists. As you identify with a community, landscape, habitat, or species, you must also grapple with its future. What will become of this land? Will its ecological integrity be maintained? Does pollution threaten community well-being? Environmentalism has emerged within the wave of rapid global environmental change, including widespread economical development, the exploitation of natural resources, the prospect of widespread pollution, and the settlement of vast areas of previously undeveloped land.

Every environmentalist must consider the extent to which he or she will become involved in working to preserve wild places and protect community health. Environmentalism, above all, calls attention to the plight of the commons—the air, water, land that all people and species share (this is the topic of chapter 3). This entails public involvement. John Muir and Rachel Carson, in particular, recognized this and grappled with the responsibilities of ecological citizenship. As eloquent spokespersons for the preservation of nature, as role models for ecological identity, they would inevitably emerge in the public eye, and great demands would be placed on their time, to speak and participate in various arenas of environmental politics. Both Muir and Carson could be described as powerful, if ambivalent, environmental activists—private people who were far more comfortable in the field than in the legislature.

Thoreau was rarely in the public eye, and he lived in a time that predates modern environmental politics. Yet he also struggled with notions of civic responsibility and public duty. Two of his essays, "On Civil Disobedience" and "Life Without Principle" are influential statements regarding the impermanence of government, the strength of individual conscience, the boundaries of citizenship, and the meaning of freedom. The time had not yet arrived for environmental advocacy, but Thoreau was a strong, if somewhat iconoclastic advocate for his way of simple living, and in doing so, he was a peculiar anomaly among what he perceived as his otherwise industrious and pecuniary neighbors. He lived during the first great wave of American industrialization, surrounded by an ebullient optimism regarding railroads, western expansion, and industrial enterprise. But in the midst of all of this, Thoreau sought his inspiration elsewhere and publicly challenged what he considered the false expectations of mindless enterprise.

The idea of simple living, although it portends withdrawal from society, can be a powerful political statement and certainly implies an approach to ecological citizenship. American environmentalism is replete with revised notions of affluence—calling attention to how people live, where their material wealth comes from, how a materialist orientation contributes to the exploitation of natural resources, and so forth. By living simply, and doing so in the public eye for clearly articulated ecological reasons, one challenges many prevailing cultural notions and makes a strong political statement.[13]

Thoreau and Muir serve as interesting role models for simple living. Muir was the prototypical backpacker, rambling for weeks at a time, requiring few meals (satisfied with oatmeal and crackers), cheerfully sleeping on beds of pine needles, finding pleasure just by watching a storm pass. He was an ascetic, rugged mountaineer, a champion of wilderness living. Thoreau demonstrated that simple living could be achieved just a few miles from town, in a small cabin, with little money, and few conveniences. The Walden Pond idea has become a symbol for conscious frugality, a challenge for generations of persons who want to cast off the trappings of material life

Yet Muir also understood that he had a responsibility to enter the corridors of traditional American politics. He became one of the first national environmentalists, fighting to preserve wild land in the form of national parks, actively trying to influence the orientation of the United States Forest Service, serving as the vanguard against real estate developers and industrialists. He understood how the national interest was defined in economic terms. To promote wilderness preservation for nonutilitarian purposes was an enormous political and moral challenge. Muir was keenly aware of the great power of American capitalism and the enormous demand for natural resources, and he wanted to counter these forces by proclaiming the importance of wilderness preservation. He had a moral obligation to speak for the wilderness.

By entering the political arena, Muir had to make numerous compromises. First, he had to spend less time in the wilderness. He knew that he could not live in the mountain forests forever, because if he did not actively promote his point of view, those forests would one day be gone. Yet he also knew that he couldn't dwell exclusively in the political realm because if he did, he would lose his mountain spirit. Second, he had to develop a language for the preservation of nature in a culture that stressed material values. Could he develop a political voice that would not be perceived as extremist or peculiar, yet at the same time be true to his values?

Muir was not only a wilderness man but one of the first and foremost environmental professionals. As such, he suffered pain and loss when his advocacy efforts failed and when utilitarian interests prevailed. But he learned how to build coalitions. And he developed a voice for preservation, explaining with strength and clarity how the protection of nature required long-term political thinking—a politics of posterity. Muir had a sense of his place in history, not because he was driven to be remembered, but because he believed so fervently in his cause and devoted his life to living it.

Several of Muir's biographers (Frederick Turner, Stephen Fox, and Michael P. Cohen) devote considerable attention to Muir the activist and Muir the wilderness man because this is such a compelling issue for contemporary environmentalists, who wonder whether such roles are compatible, and how they can be effectively balanced. Muir wandered back and forth, from the high adventure of his wilderness travels, to the domestic landscapes of his family and business, to the complex domain of national politics.[14] He sought (not always successfully) to balance solitude and companionship, the wilderness and the city, the wild and the domestic, spontaneity and discipline, spiritual development and political responsibility. For years before his destiny unfolded, he wandered without a sense of profession or vocation. He spent years developing a voice for his moral concerns—struggling to determine where he belonged, and what responsibilities he had to promote his strong point of view. It is revealing to juxtapose four images of Muir: climbing a tree in the middle of a snowstorm, camping out in the Sierras with Theodore Roosevelt (more about that later), testifying before Congress, and sitting at home with his wife and two daughters. He may not have always reconciled these different roles, but he demonstrated the importance of each of them.

Rachel Carson's great love was field ecology because this was the way she learned how to celebrate life. As a popular science writer and an emerging literary figure, she understood how to convey her enthusiasm and sense of wonder to a large audience. It was these qualities—her reputation as a writer, her dedication to life, her love of natural history—that eventually made her a compelling public figure. In 1962, she finished *Silent Spring,* widely regarded as one of the most important books in the history of environmentalism.

In contrast to her previous books, *Silent Spring* had a very dark message, as it warned the American public about the pervasive, insidious, and sinister prospects of the widespread use of pesticides. As her biographer Paul Brooks indicates, *Silent Spring* involved a different kind of research. Rather than the delights of special places, she was now researching the dire threats to those places. "Joy in the subject itself had to be replaced by a sense of almost religious dedication, and exhilaration in searching out the truth which sustained the author through extraordinary trials."[15]

Rachel Carson was now entering the toxic world of DDT and other pesticides. And this was groundbreaking research. Few people had questioned the scientific and agricultural establishments' near-unanimous support of pesticide use. Her work took extraordinary courage for four reasons: she had to become fully immersed in the ecology of love and loss, she had to collect enough research to challenge the prevailing view regarding pesticides, she would suffer the indignities of a woman in the public eye who was playing a role that had typically been reserved for men, and she herself was suffering from declining physical health.[16]

The full story of *Silent Spring* has been told elsewhere and the interested reader should read Paul Brooks's *The House of Life*. What is most important is how Carson's research became intrinsic to her sense of ecological identity. When she became aware of the threats to life that were represented by pesticides, she had no choice but to enter the public realm, using her strength as a nature writer, and her credibility as a "science writer." She understood that pesticides were merely one aspect of a much larger pollution problem. So it was her moral drive, cultivated through her ecological identity, that inspired this new direction. And during the most difficult phases of her work, when she faced enormous pressure and public scrutiny, she found support and solace through occasional glimpses of the natural world, whether it was spotting a formation of geese through an open window, or a simple walk outdoors.

Although Carson was widely criticized by the scientific establishment, she received enormous grassroots support, and eventually other studies vindicated her assessment. Through her perseverance and courage, she became a role model for ecological citizenship, demonstrating how grassroots organization, impeccable research performed by both amateurs and experts, and female leadership could have a major impact on public policy. These themes have been crucial to the development of contemporary environmentalism. And finally, Rachel Carson's work reiterates that the formation of ecological identity inevitably involves an ecology of love and loss, and each person must find a way to confront this issue.

Individuals like Henry David Thoreau, John Muir, Rachel Carson, and countless other environmental heroes serve as the conscience of environmentalism. If their perspective appeared unyielding at times, it is because they saw no alternative. They were motivated by their ecological identity. Their search to find themselves in nature gave them the fortitude and courage to pursue their goals, even when they took positions that alienated mainstream thinkers. Their ecological identity was the source of their vision and provided their clarity of purpose, serving as the moral anchor for difficult political choices. In this way, they inspire the broad range of the environmental spectrum and reveal a complex network of branches on a collective environmental tree.

The practice of the wild, the natural history excursion, the path of ecological citizenship, these constitute the roots of the American environmental tradition. They are means for the expression and exploration of ecological identity. As I reflect on the great number of "trees of environmentalism" that have passed through my class, I find that these themes have become a recurrent pattern. People want to explore these ideas in depth—how to apply them to everyday life and career decisions, how they can be used to inspire community service and political action, and how they provide a moral and spiritual outlook. Yet as these ideas move through the branches and leaves of the trees, they may take different forms. In the branches, some of the controversies of environmentalism begin to emerge. People may concur in their enjoyment of wild places, but they may have vastly different ideas about how those places should be managed. They may agree as to the necessity of environmental politics, but their strategic approach or their evaluation of broader political questions may be diametrically opposed. In the next two sections, some of the issues behind these controversies begin to unfold.

Branches: The Environmental Spectrum

Preservation or Conservation?

In 1903, John Muir and Theodore Roosevelt spent 4 days camping together in Yosemite National Park. By all accounts they greatly enjoyed each other's company. Muir found a challenging comrade: Roosevelt was loquacious, boisterous, and confident. Moreover, he was as enamored with natural history as Muir, pointing out birds that escaped Muir's attention. He shared Muir's enthusiasm for wild places and the simple life, sleeping cheerfully on the evergreen-bough bed which Muir prepared for him. To the dismay of Roosevelt's entourage (including an aggressive press) the two men wandered through the backcountry, avoiding all human contact, and determining their own agenda.

Why did Roosevelt camp out with Muir in the first place? And why would Muir, the staunch defender of wild lands, a man who distrusted professional politicians, consent to spend time with the President? Muir and Roosevelt both loved the wilderness. Many of Roosevelt's formative years were spent in the wilderness. Like Muir, it was in the wilderness that Roosevelt gained his confidence, manhood, and personal strength. He appreciated wild places and he wanted to meet the man who had done so much to proclaim their importance. Muir recognized the political potential of such a meeting, knowing that he might convince the President to promote the preservation of more Sierran forests, hoping perhaps that he could appeal to the President based on their shared enthusiasm for the wilderness.

Muir and Roosevelt understood the historical context of their encounter. America's wild lands were rapidly disappearing and the pressures of lumbering, mining, and real estate were impinging on many of America's most sacred places. It was imperative, they agreed, to find some way to preserve these places because without them the soul of America would vanish, overtaken by the engines of industrialism, subdued by rampant commercialism. Yet Roosevelt and Muir had vastly different wilderness philosophies. Muir had an ecocentric orientation, seeing humans as just one of many species in a broad, cosmic context—believing in preserving wild lands for the sake of their wildness, without any ulterior human motives. Roosevelt was anthropocentrically oriented, proclaiming the importance of human affairs, understanding that resources must be simultaneously used and preserved, relying on human stewardship and wisdom to achieve that balance. Although they shared a love of the outdoors and commonly understood the critical threats to wilderness lands, their philosophical and political orientations were significantly different.

Muir and Roosevelt stood together in the High Sierras and peered into the American future. They were grappling with the same questions that would plague environmentalists in decades to come. Muir appealed to the heart; he understood that a preservationist perspective could best be derived from experience and that to appreciate the importance of wilderness a person had to identify with nature, communicate with nature on the most profound spiritual level. Roosevelt, despite his love of the wilderness, demonstrated the importance of the practical side, showing how by appealing to the utilitarian and economic contingencies of wise use, one could forge a consensual approach to stewardship, conserving resources for human use while preserving the most sacred places. Although wilderness protection is important, economic welfare is crucial, and there are many circumstances in which the demands of a burgeoning economy outweigh those of wilderness preservation.

Should both Muir and Roosevelt be considered environmentalists? Muir was the heart and soul of preservationism, proclaiming the moral significance of wilderness. Roosevelt was the epitome of conservationism, a man who could love the outdoors, and dedicate his presidency to the stewardship of natural resources. He spearheaded a century of environmental legislation that was responsible for securing vast tracts of national forests.

Some environmentalists contend that the national forests were not sufficiently preserved, that economic development has always been intrinsic to the mission of the forest service. Others argue that although the national forests allowed some development and were not wholly preserved, what more could one expect given the national predilection for laissez-faire economics and rapid commercial growth? This controversy is fundamental to American environmentalism.[17] The preservationist and conservationist positions are often incompatible, and in many situations involving environmental politics, the preservationist point of view is compromised by what is perceived as the pragmatic, conservationist approach. In effect, Muir and Roosevelt set the stage for a critical theme of American environmental politics.[18]

Many environmentalists find these positions mutually exclusive. Preservationists consider the language of costs and benefits a bald compromise, a perspective that completes a self-fulfilling prophecy, the eventual destruction of the wilderness, doomed by the primacy of economic growth. According to this view, environmentalism must present Americans with a stark reality: only a fundamental revision of our approach to nature will enable us to develop the ethical and moral commitment to preserve wild land, and only such an approach will prevent catastrophic ecological destruction. We must see ourselves for who we are, an imperial species with scant regard for other life forms. To restore the earth, we must reevaluate our approach to nature. Conservationists find this position extreme, emphasizing that humans have the managerial skill and technical ability to serve as the custodians of the earth's resources. This is an instrumental imperative; the techniques of scientific ecology provide us with the knowledge and the means to balance economic growth and environmental quality. People don't need to change their worldview as much as they have to regulate their behavior and develop policies and procedures that enable Americans to efficiently manage their resources.

About 70 years after Muir and Roosevelt's camping trip to Yosemite, John McPhee wrote *Encounters with the Archdruid,* a book that vividly raised the preservation/conservation controversy. McPhee arranged a series of wilderness encounters for David Brower, whom many consider to be the preeminent twentieth-century heir to Muir's legacy. Brower is the "archdruid," a diehard preservationist, who spends his life defending the wilderness, alternately roaming high mountains and barnstorming the country, speaking out, planning political strategies, working diligently, and at times outrageously, for the preservationist cause. Few contemporary figures have been more steadfast in their approach to environmental politics or more outspoken in their wilderness philosophy.

His encounters are with men who have diametrically opposed perspectives on wilderness use. McPhee arranges for Brower to hike in the Cascades with Charles Park, a mineral engineer; to visit Cumberland Island with Charles Fraser, a real estate developer, and to ride down the Colorado River on a raft with Floyd Dominy, the infamous dam-building federal bureaucrat. In each case, we wonder whether Brower can defend his wilderness philosophy against the strong utilitarian perspective espoused by his fellow travelers. What emerges is a complex portrait of the issues that divide these men—and the preservationist and conservationist perspectives.

Charles Park is a member of the American Ornithologists' Union. McPhee attests that he has never spent time with anyone who was more knowledgeable about the natural world, describing him as a "man who knows what he is looking at in wild country."[19] Park has toured the world's wildest places in search of mineral treasures. He derives enormous pleasure from his wilderness adventures, yet his overriding purpose is to find deposits of valuable minerals so that they can be extracted and exploited for commercial purposes. Brower and Park, after a long conversation that takes place during a strenuous hike, pause to enjoy a cup of wild blueberries. Brower remarks that he feels sorry for all of the people who just don't understand what mountains are good for. McPhee asks what they are good for. Brower says blueberries. Park says copper.

Park is dismayed at Brower's hypocrisy, claiming that Brower's life would not be possible without the minerals he takes for granted. He asks Brower where the chrome on his refrigerator door comes from, the nails that hold his house together, the pigments in paint, or the tungsten in light bulbs. Brower claims that he would rather do without these minerals if it meant preserving wilderness and promoting ecological diversity. But Park has no patience for that perspective, finding it naive and privileged. Brower replies that environmentalism is about changing the way we live, understanding the impact of our actions, placing the human species in an evolutionary and ecological perspective.

Fraser, the developer, and Brower contemplate the recreational use of wilderness. Fraser wants to generate income for his development plans in a way compatible with a preservationist perspective. Without ever admitting so, he desperately wants Brower's approval. Brower plays the sage. He urges Fraser to persuade other realtors that they have an environmental responsibility. The reader wonders under what circumstances Brower would approve of Fraser's development plans. Both men are coy; they use each other to legitimate their positions, but we wonder whether it's merely an uncomfortable compromise. Brower admits that there is a place for recreational development. Fraser wants a preservationist to approve of his plans.

Dominy and Brower, however, seem to be irreconcilably antagonistic. Dominy revels in the damming of the Colorado River, proclaiming Lake Powell an economic and recreational bonanza, citing the number of people who now have access to the region, and who have benefited from the dam. A western folk hero, Floyd Dominy overcame a harsh economic childhood in rural Wyoming to become a federal lands manager, bringing dams, recreation, and prosperity to large portions of the American West. In contrast, Brower laments the flooding of this once-exquisite landscape, whose uniquely sculptured canyons are now forever hidden from view. Dominy thinks Brower is selfish to deny access to these areas and that Brower will only be satisfied once every piece of the American landscape is placed under lock and key. Brower, in turn, views Dominy as rapacious and unquenchable in his effort to develop all of the American West, without regard for ecological consequences, and without appreciation of wilderness aesthetics. Neither man blinks in the encounter. They share a great joy in riding the river. But we wonder whether this thread is enough for them to find some common ground. The tension between Dominy and Brower remains unresolved.

Typical of conservationists, Park, Fraser, and Dominy aim to utilize natural resources wisely and efficiently. They recognize the importance of preserving some wild lands, but they are mainly concerned with how humans can best use those lands. They are not necessarily unthinking developers, rather they view the preservationist agenda as impractical and elitist. Their encounters with David Brower recapitulate the experience of Muir and Roosevelt. The contrasting perspectives, although representing ideal types, seem to reflect a dynamic tension of contemporary environmentalism.

Often these positions are portrayed as a choice between economic growth and environmental quality, or between jobs and wilderness. When environmental issues are discussed in the public arena, these stereotypes are likely to emerge. If environmentalists speak too fervently for land preservation, they may be perceived as elitist and opposed to economic growth. To what extent must people compromise a strong preservationist orientation, if they are to be taken seriously?

Conservationists espouse what they perceive as a more practical position, the best one can hope for given the American ideological predisposition regarding economic growth.

Many environmentalists are influenced by this preservation/conservation dynamic. Romantically, they are strongly persuaded by people like Brower and Muir, inclined to preserve all wilderness, deeply committed to pursuing a way of life that conforms with this approach. Yet practically, they are solidly embedded in a consumer economy, taking pleasure and profit in the fruits of the exploitation of the earth. Is this hypocritical? Is the preservationist perspective merely a false image of what many environmentalists imagine themselves to be? Or perhaps this distinction itself is superficial, creating ideological dichotomies that can be transcended once an attempt is made to cut through the stereotypes. For example, the choice between jobs and wilderness may only reflect the way particular interest groups or the media portray stereotyped positions, posing unnecessary choices, effectively placing environmentalists and workers in opposite camps, when they really have a great deal in common—the long-term health of their community.

Controversies such as this indicate that there is a great deal of political and philosophical diversity among environmentalists. Ecological identity work entails understanding the history of American environmentalism, so people appreciate the full spectrum of positions and possibilities, recognizing the historical and social context of environmental choices. Environmental beliefs are grounded in a cultural milieu, lodged in politics and economics, embedded in questions of public policy as well as ecological identity. Most environmentalists, similar to Brower, must understand the political dimension of their work—the public expression of ecological identity, and be willing to deal with the controversial ramifications (see chapters 3 and 4).

Beyond the Spectrum

In much of the environmental literature, preservationism and conservationism are presented as an entangled spiral, a double helix, each claiming to represent the mainstream of American environmentalism. Many histories of the environmental movement portray the dynamic between these positions, offering various interpretations of the relationship between people, ideas, and organizations from this interpretive perspective. But there are other distinctions as well, and although the preservationist/conservationist dynamic is important, there are alternative ways to conceive of the environmental spectrum.

For example, Donald Worster, in *Nature's Economy*, shows how the intellectual history of ecological ideas is crucial to understanding modern environmental thought. One root of ecological thinking lies in the romantic, transcendental, organic, vitalistic, holistic realm. Another root is embedded in the analytical, scientific, mechanical, and utilitarian realm. Worster illustrates this divergence by discussing the work of Thoreau, Darwin, and Aldo Leopold, among others, and explains how ecological ideas have reflected a dynamic tension between these approaches, how these men often embodied both traditions in their approach to knowledge.[20] This distinction reiterates the cognitive and intuitive aspects of understanding ecology, as discussed in chapter 1.

Stephen Fox in *The American Conservation Movement* reflects on the difference between amateur and professional environmentalists. The amateurs are grassroots-oriented, motivated by strong values, antimodernist in sentiment, unconventional, strongly influenced by the models set by Thoreau and Muir. They aspire to organize their local communities and often pursue alternative lifestyles. The professionals are urban, institutional, normative, establishment, strongly influenced by consensual public policy models, and are more likely to work for mainstream organizations. Peter Borelli, in *Environmentalism at a Crossroads,* describes environmental groups as either norm-driven or value-driven, noting how such dilemmas are intrinsic to most decision making. Norm-driven groups are more likely to work within the system, to make compromises, to use mainstream institutional channels to accomplish their objectives. Value-driven people, according to Borelli, will orient their life choices around an environmental perspective; they are less willing to compromise, more willing to act on the basis of their ecological identity.

In *Forcing the Spring,* Robert Gottlieb suggests that many of these interpretations exclude several important elements: people who work in the various public health fields combating the effects of industrial pollution and urban decay, and people who encounter toxic environmental hazards. Gottlieb links the workplace health and safety movement and the labor movement to environmental concerns. He describes the importance of environmental justice—how people of color are challenged by discrimination, exclusion from power, and are also likely to be subjected to community environmental hazards. Gottlieb tells the story of another stream of heroes, historical figures such as Alice Hamilton, "the mother of American occupational and community health";[21] or Rachel Carson, who had to withstand outrageous anticommunist and sexist innuendoes in defending herself against American industry; and contemporary figures such as Penny Newman, "who organized her own community of Glen Avon to fight against contamination seeping from the nearby Stringfellow Acid Pits east of metropolitan Los Angeles." [22]

In his introduction, Gottlieb describes the first national People of Color Environmental Leadership Summit, which included a diverse group of grassroots environmental activists. He cites a speech by Dana Alston, a key organizer of the summit:

> Our vision of the environment is woven into an overall framework of social, racial, and economic justice. The environment, for us, is where we live, where we work, and where we play. The environment affords us the platform to address the critical issues of our time: questions of militarism and defense policy; religious freedom, cultural survival; energy-sustainable development; the future of our cities, transportation; housing; land and sovereignty rights; self-determination; employment- and we can go on and on.[23]

The literature of environmentalism abounds with interpretive taxonomies, various classification schemes, and approaches that attempt to place environmentalism in social, intellectual, and historical contexts. Such interpretations are helpful in that they illuminate the origins and diversity of environmentalism, pointing out some of the controversies within the branches of the environmental tree. They provide a structure and a collection of interpretations that help people visualize the meaning and ramifications of environmentalism. Collectively, these scholarly interpretations reveal a broad spectrum of contemporary environmental thought.

The idea of an environmental spectrum serves as the basis for an interesting learning activity, what I call "beyond the spectrum." I use this approach in two contexts: as a precursor to the environmental tree (offering people some ideas about how they might organize their knowledge of environmentalism) and also as an introductory activity for diverse groups of environmental practitioners. After a discussion about environmentalism's broad and rich tradition, I suggest to a group that we spend an hour trying to collectively organize the various strands of that tradition. I propose several axes as a way to classify the material: preservation/conservation, ecocentric/anthropocentric, grassroots/professional, radical/normative, and so on. The group decides which of these distinctions, if any, they find useful, and whether any clarifications or modifications are necessary. After some discussion, I enter the relevant distinctions on opposite ends of a continuum on a blackboard or newsprint.

We engage in a collective brainstorming. The group names any people, ideas, or organizations that are important to environmentalism. I ask the group to instruct me as to where each name belongs on the chart. In some cases, there is little controversy. For example, EarthFirst! clearly belongs at the ecocentric, grassroots, radical end of the spectrum. In other cases, however, there is a great deal of ambiguity or controversy. An idea such as bioregionalism crosses several categories. An organization such as the Sierra Club has members from all over the spectrum. When this occurs, I draw a line indicating the span and reach of the concept or organization. Eventually the diagram is complete—the group runs out of names or the chart is filled. A complex collage forms.

I ask the participants to imagine that our room represents the diagram. This calls for a degree of simplification, such as highlighting the most pronounced distinctions. For example, one corner of the room may depict the preservationist extreme, the other corner the conservationist extreme. Each person moves to the place in the room that best describes where he or she belongs in the spectrum of environmentalism. But I require two sequential placements: the first time according to what the heart suggests (which approach most reflects your values?) the second time according to the professional persona (which approach best describes the work that you do?). A period of confusion typically ensues. I entertain various protests: people are not ready to commit themselves, they don't know enough yet, they're in the class to find out where they belong, it's superficial to create stereotypes, these are limiting distinctions. I suggest that they are merely avoiding the issue, that pressing decisions come up all the time, and that if they look deeply into themselves, their predilection should be clear.

Several interesting patterns emerge. First, it is evident that in spite of some initial ambivalence, most of the participants eventually realize that they have strong convictions and they move to the corresponding place in the room. Second, there is typically a great deal of movement when they must realign themselves according to their professional persona. This movement is almost always toward the middle, from either end of the spectrum. The large majority of my students have a strong preservationist orientation, but many work for organizations or exhibit a professional persona that reflects a more moderate approach.

Some students find that their professional work is perfectly aligned with their values and they do not move at all, but this is less often the case. Finally, some students are frustrated by these dichotomies, feeling that there are other streams of environmentalism, that they are really not comfortable describing themselves in this way.

For example, after his frustration with the limitations of traditional interpretations, one student offered a more inclusive taxonomy built around three interlocking streams of the environmental tradition: wilderness preservation and respect for nature; resource conservation and sustainable economics; and public health and environmental justice. He suggested that these categories allow environmentalists to break through the traditional preservation/conservation dichotomy and to understand the breadth of environmentalism as both a social movement, a philosophy of nature, and a search for personal meaning.

This activity teaches a great deal about the breadth and diversity of the American environmental tradition. Collectively, students figure out where they stand, what's important to them, and what orientation they follow. They also recognize that regardless of their place in the spectrum, all aspects of the environmental tradition may be embodied in their work. Their understanding of ecology undoubtedly comprises both the romantic and analytical traditions. As environmental professionals or concerned citizens, they have often worked in both grassroots situations and larger institutional settings. At times people are simultaneously norm-driven and value-driven. Often people move through radical and normative approaches, depending on the circumstances of their work.

In formulating an ecological identity, people widen their circles of identification, not only in terms of the direct experience of nature but also within a professional and political context: choosing organizations to belong to, selecting forums for political action, defining professional commitments and orientations. To say that you are a member of Greenpeace, or that you work for a conservation commission, or that you are heavily influenced by David Brower enables you to declare affiliations, not in an exclusive or individualistic sense, but so you can find yourself in the spectrum, and distinguish yourself accordingly. Environmentalism brings people into the political and professional arena, just as surely as it helps them understand their relationship to nature.

This activity helps people realize that although they are forging a distinct path, they have the benefit of a rich tradition with diverse perspectives. Within their careers and life choices, they feel the tug of all of these streams; they can be enriched and influenced by all of them, yet still act on their deepest convictions.

Muir and Roosevelt anticipated a century of environmental controversy, and their positions defined some parameters of the environmental debate. However, 100 years later, the ecological imperative is increasingly urgent. The ominous prospects of global ecological change portend impacts that Muir and Roosevelt could barely have contemplated. For example, Edward O. Wilson, in *The Diversity of Life,* explains that the greatest threat to the planet is the decline in species diversity because this disrupts the very fabric of ecosystems and ultimately threatens the human species.[24] Large ecosystems cannot thrive unless they are preserved intact. Hence the environmental agenda is more than a function of aesthetic and spiritual inclination, it is fundamental to human survival. As the recognition of environmental decline becomes more widespread—loss of habitats, species, sacred places, and ecosystems—environmentalists will more fervently search to construct interpretations that integrate political, economic, and social perspectives with the psychological and spiritual process of widening the sense of self in respect to nature. In the next section, I survey several approaches within environmentalism that illuminate the concept of ecological identity.

Leaves: Ecofeminism, Deep Ecology, and Bioregionalism

The Environmental Trees Have Thick Canopies

In the last 25 years the environmental movement has exploded. Mainstream environmental organizations such as the National Audubon Society, the Sierra Club, and many others have dramatically increased their membership. Many environmental groups have become large organizations, involved in business, advocacy, education, litigation, and even real estate, with offices in several cities, and international branches. The environmental movement grows increasingly complex, taking on multiple agendas, moving to the center of public life.[25]

Browse through any reasonably sized newsstand and you find numerous environmental magazines, not to mention professional journals and organizational newsletters. In the 1960s, there were no explicitly environmental wide-circulation magazines, and only a few specialized ecology journals. Contemporary bookstores typically have large environmental sections, filled with field guides, nature writing, books about environmental politics, books about saving the earth, and so forth.

In the late 1960s, there were only a handful of undergraduate and graduate programs in environmental studies. These fledgling programs had to fight recalcitrant administrators and staunch academics who wondered about the feasibility of such an interdisciplinary approach. Now, in the 1990s there are hundreds of such programs. There are book-length guides to environmental studies programs and careers. On college campuses, environmental studies courses are hugely popular. At least a half-dozen career newsletters are available, listing jobs, internships, and other notices about the environmental profession.

The extraordinary growth of the environmental movement has been accompanied by a dazzling array of ideas, including research in ecology, the emergence of various interdisciplinary approaches, and the synthesis of interdisciplinary methodologies. This has spawned new literature, philosophies, and language. Consider just a few of the interesting words and concepts that have emerged in the last 25 years: sustainability, biodiversity, ecological restoration, bioregionalism, deep ecology, ecosophy, ecofeminism, social ecology, the Gaia hypothesis, global environmental change, Green political thought, ecological economics. It would take several large volumes just to survey the development and evolution of these ideas.[26]

These ideas, concepts, people, and organizations make up the leaves of the environmental tree. In my class, when the students view one another's trees, they are easily overwhelmed by the scope of the material. Environmentalism turns out to be far more complex than they ever imagined. There appear to be so many ways that the idea of environmentalism can penetrate their lives. People wonder which ideas are most useful to know about, which can provide them with the most guidance and insight. I explain that one cannot study and explore all of these approaches, although it is useful to know about them. Rather they can focus on whatever ideas they find most appealing. It is enough to know where the leaf fits on the tree, but it is instructive to study the leaf in closer detail.

In one way or another many of these leaves represent expressions of ecological identity. If a person chooses a career in ecological restoration, he or she will deal with the science of reconstructing disturbed habitats. Inevitably, such work involves identification with nature. If a person is aroused by the idea of sustainability, he or she will study and apply an approach to natural resource use that emphasizes, among other things, ecological criteria as the basis for economic development. Yet sustainability is also a powerful metaphor which informs ecological identity—the quality of life is a function of ecosystem health and viability. One cannot make such a conceptual leap without widening the sense of self in relationship to nature.

In this section, I review ecofeminism, deep ecology, and bioregionalism because they have so much educational potential for the personal and political expression of ecological identity. These concepts have had an enormous influence on the way people think about environmentalism. Although they are not household words, and are often associated with the more radical, controversial wing of environmentalism, they address how people experience nature and serve to challenge many prevailing notions about education, personal behavior, and political action. As I peruse the environmental trees in my classroom, I discover that these are some of the concepts that people are most interested in and excited to talk about.

Ecofeminism

Ecofeminism represents a wide and diverse body of theory and practice. The word connotes the integration of feminism and ecology. Hence the integrating theme throughout ecofeminism is the reexamination of the relationships between women and men and between humans and nature. In *Radical Ecology*, an excellent review of contemporary environmentalism, Carolyn Merchant describes the various approaches to the concept, identifying liberal ecofeminism, cultural ecofeminism, social ecofeminism, and socialist ecofeminism. These groups represent vastly different political philosophies, theories of nature, theories of human nature, and recommendations for personal action. So it is crucial not to make sweeping generalizations about what amounts to a complex, emerging literature.

Charlene Spretnak, in an important essay, "Ecofeminism: Our Roots and Flowering," describes three paths of entry to ecofeminism, each representing an "occasion for awakening."[27] First there is the realization that the dominance of male over female "is the key to comprehending every expression of patriarchal culture with its hierarchical, militaristic, mechanistic, industrialist forms." This domination subverts the expression of female identity, produces exploitive images of both women and nature, and devalues the very process of connecting to nature. Many women find a revealing parallel between how humans treat nature and how culture treats women. This realization emerges both from personal experience and exposure to the literature of political theory and history.

The second path to ecofeminism is "exposure to nature-based religion, usually that of the Goddess." Spretnak refers to a religious awakening, inspired by "rituals of our own creation that express our deepest feelings of a spirituality infused with ecological wisdom and wholeness." Women create an ecospirituality based on their experience of the earth, their connections to the land, their ability to identify their bodies with the whole planet—developing metaphors of cyclic renewal, regeneration, and fertility. Ecofeminism generates a new language of the sacred, emerging from these metaphors. "What was intriguing was the sacred link between the Goddess in her many guises and totemic animals and plants, sacred groves, and womblike caves, in the moon-rhythm blood of menses, the ecstatic dance—the experience of *knowing* Gaia, her voluptuous contours and fertile plains, her flowing waters that give life, her animal teachers."[28] This approach has had a profound impact on ecospirituality, influencing the mainstream religions, legitimating the exploration of alternative psychological and spiritual paths.

The third path is environmentalism itself. Female environmental professionals discover that their career path is blocked and feminism provides a framework for understanding why this is so and what can be done about it. Women and men who become involved in environmental politics realize that the problems they encounter transcend simple explanations—they require a more systematic social analysis, leading to explorations of power and exploitation, connecting them to feminist theory. Students in environmental studies classes find that they can combine their interests in nature and feminism. Their concept of personal identity is enriched and challenged by combining their experience of nature with their experience of gender.

It is impossible to reflect on the full scope of ecofeminism in this space. Needless to say it has complex and deep philosophical, psychological, and political implications and is the source of a great deal of controversy both within itself and among environmentalists. What is important here is its educational ramifications for ecological identity. Ecofeminism provides some people with an academic and experiential home, a way for them to comprehend their experience both as environmentalists and as women or men. The interface of feminism and environmentalism allows people to assess this experience, offering a framework for the exploration of personal identity.

For example, research in feminist psychology is attempting to reevaluate male-based perspectives on human development which focus on independence, individualism, and autonomy.[29] The ecofeminist orientation emphasizes the cultivation of interdependence, relational ability, belonging, nurturing, sensitivity and intuitive ability. Can an ecofeminist perspective on child rearing and adolescence provide insight regarding the emergence of an ecological worldview? If so, this has extraordinary ramifications for environmental education, parenting, and personal growth. The practitioner can construct experiences which facilitate the formation of ecological identity: incorporating ecologically based rituals in religious holidays and everyday life, developing rites of passage based on seasonal cycles and the diversity of life, using forms of artistic expression to access being *in* nature rather than always learning *about* nature, using storytelling to reweave one's experience as a relationship to the earth, organizing children's play to emphasize the magic of nature.[30]

One of my students, Kristin, in explaining the place of ecofeminism on her environmental tree, reflects on how the ecofeminist approach informs her search for ecological identity and her orientation as an environmental educator:

> Ecofeminism aims to heal the separation between female and male experiences and values. In an ecofeminist worldview, feminine categories are advocated as equally valuable ways of seeing and structuring the world. Both sexes are encouraged to develop their emotional sensibilities, intuition, and nurturing qualities as equal complements to reason, rationality and objectivity. If education teaches people ways of knowing, then ecofeminism as education is concerned with validating a different way of knowing outside of the male-defined experience. An ecofeminist education assumes that humans as an interconnected part of nature have knowledge about themselves that is very different from scientific, rational, book knowledge. By accepting our interconnectedness, by internalizing that we are nature, human beings can understand how they are integral parts of the human-earth community. Ecofeminism helps educate people in a way that encourages them to sustain and nurture a community of diverse but interconnected forms of life.

For Kristin, ecofeminism becomes the transcending theme of her environmental tree, providing a belief system that informs her interpretation of environmentalism. In cultivating her vision as an environmental educator, Kristin aspires to apply ecofeminist approaches to curriculum development and classroom teaching.

Deep Ecology

Similar to ecofeminism, deep ecology represents a wide, diverse, and often controversial emerging literature. "Deep" connotes an attempt to uncover the most profound level of human-nature relationships, stressing the need for personal realization as accomplished by integrating the self with nature. The central theme of deep ecology is the fundamental reevaluation of conventional, modernist perspectives on the role of humans in the natural world, and an orientation that stresses that all life on earth has intrinsic value, that the richness and diversity of life itself has value, and that human life is privileged only to the extent of satisfying vital needs. In *The Idea of Wilderness*, Max Oeschlaeger emphasizes that deep ecology is committed to an explicitly ecocentric orientation and constructs alternative social ideals and values to that end.[31]

People come to deep ecology from three interconnected paths. First is the perspective of radical environmental politics. Deep ecology opposes what it calls "shallow ecology," or promoting reform through normative institutional frameworks—the conventional channels of environmental legislation, representative democracy, and traditional environmental education. Rather, it asserts that solving the environmental crisis requires a radical new vision, including a fundamental reassessment of consumption and production, basic changes in the economic and political structures of advanced industrial civilization, a reorientation of how quality of life is assessed, and an emphasis on local autonomy and decentralization. Deep ecology has significantly influenced green political theory and radical activism, although it should also be mentioned that there is significant controversy within radical environmentalism as to ends and means, theories of the state, and approaches to human-nature relationships. Deep ecology is merely one of several approaches to radical environmentalism.[32]

Deep ecology also emerges from a philosophical perspective, including a critique of conventional notions of scientific progress and modernist philosophy. Much of the deep ecology literature is an attempt to synthesize a philosophy of nature that has roots in Native American wisdom, Oriental philosophy, radical anthropology, romantic ecology, wilderness poetry, and numerous other sources. Deep ecology emphasizes a reconstruction of self based on Arne Naess's notions of wider circles of identification (see chapter 1) and an ecocentric perspective on environmental ethics—the intrinsic value of all life forms. This has spawned a dense, controversial, and often esoteric philosophical literature, as well as an ecopsychological perspective.

There is a third, practical approach to deep ecology which emphasizes personal behavior, community living, and environmental education. For example, the classic treatise of Bill Devall and George Sessions, *Deep Ecology: Living as if Nature Mattered,* serves as a guidebook to the personal, political, and educational applications of deep ecology. The point of this book and of subsequent books by Bill Devall is that one has to live deep ecology by cultivating a relationship to the earth based on experiences in nature, consumer behavior, and critical self-reflection. Deep ecology is fundamentally concerned with what people can and should do to live a green lifestyle.

Deep ecology has an environmental education wing, which constructs a curriculum, offers teacher workshops, and organizes approaches to learning that explore the formulation of ecological identity. The Institute for Deep Ecology Education runs an annual retreat, including prominent practitioners and academics, with the explicit intent of integrating deep ecology with personal growth and professional vision. Some of the educational activities include naturalist studies, "listening" to trees, restoration ecology, the built environment, ecopsychology, and evaluating model environmental education approaches. Bill Devall describes this as applied deep ecology. What distinguishes this approach is the emphasis on showing how these learning approaches expand the sense of self in relation to nature.[33]

Both deep ecology and ecofeminism are easily stereotyped and relegated to the radical fringe of environmentalism. Yet a comprehensive exploration of the literature doesn't reveal a uniform perspective as much as an emerging philosophy comprised of diverse orientations. One may adhere to many of the interesting ideas in these concepts without joining a party platform, or for that matter taking on the deep ecologist or ecofeminist label. Both approaches have significantly influenced contemporary environmentalism because they have challenged so many of its assumptions. By emphasizing critical self-reflection, a critique of normative institutions, innovative approaches to interdisciplinary scholarship, and experiential learning about ecological identity, ecofeminism and deep ecology reflect an intellectual vitality that inspires many environmentalists.[34]

Bioregionalism

In the Winter of 1981, *Coevolution Quarterly 32* (now *Whole Earth Review)* ran a special issue on bioregions. On the first page was a quiz called "Where You At?" described as "a self-scoring test on basic environmental perception of place." It consisted of twenty questions, all addressing areas of fundamental environmental knowledge. Readers were asked to trace water from precipitation to tap, name five edible plants in their region, name five resident and migratory birds, describe where their garbage goes, describe the land use history of their region, and so on.[35]

The lead article of that issue, "Living by Life" by Jim Dodge, explored bioregionalism, thereby placing the quiz in a broader context, attaching it to theory and practice. Dodge explains at the outset that he is not all that sure what bioregionalism is, that it is a loose and amorphous formulation—more appropriately called a notion—rather than a formal theory. But he explains that it "has been the animating cultural principle through ninety-nine per cent of human history and is as least as old as consciousness." He follows with an etymological introduction:

> Bioregionalism is from the Greek bios (life) and the French région (region), itself from the Latin regia (territory) and earlier, regere (to rule or govern). Etymologically, bioregionalism means life territory, place of life, or perhaps by reckless extension, government by life. If you can't imagine that government by life would be at least 40 billion times better than government by the Reagan administration, or Mobil Oil, or any other distant powerful monolith, than your heart is probably no bigger than a prune pit and you won't have much sympathy for what follows.[36]

Dodge explains how a "central element of bioregionalism is the importance given to natural systems . . . both as the source of physical nutrition and as the body of metaphors from which our spirits draw sustenance." It also involves aspects of cultural behavior such as subsistence techniques and ceremonies. But most important, it demonstrates the interpenetration of cultural and natural systems. "To understand natural systems is to begin an understanding of the self. . . . When we destroy a river, we increase our thirst, ruin the beauty of free-flowing water, forsake the meat and spirit of the salmon, and lose a little bit of our souls."[37]

The basic premise of bioregionalism is that ecological considerations should determine cultural, political, and economic boundaries. Although it is unclear how such boundaries could be formally delineated (based on watersheds? landforms? species composition? psychophysical influence?), it is instructive to experiment with various criteria. The important concept is that the local ecology should determine the political economy. For example, Vermont and New Hampshire are divided by the Connecticut River, but as a watershed, the river and its ecology integrate the people who live along it. Yet they live in different jurisdictions. Many political designations are a function of imperial land use patterns rather than ecological considerations.

For Dodge, bioregionalism implies a confederation of small-scale communities that are interdependently self-reliant, committed to social equity, engaged in participatory decision making, subsistence-oriented, and responsive to environmental contingencies. These communities will develop unique cultural specializations and engage in bioregional trade. Following the work of Lewis Mumford, E.F. Schumacher, and Leopold Kohr, the emphasis here is on decentralization and appropriate scale, with the underlying assumption that environmental quality, ecological restoration, and social justice are most likely to be implemented in a community of face-to-face interactions, in a community that lives close to the earth.[38]

Above all, bioregionalism asks people to notice where things come from, how the basic patterns of everyday life are embedded in an ecological web. When you shop at a supermarket, or drive your car, or use appliances, you are likely to forget that all of the commodities you use are extracted from nature, and have an ecological and social impact. Where did that kiwi fruit come from, anyway? What about the computer I'm using to write this book? Bioregionalists advocate regional self-reliance, using local resources whenever possible; making local decisions; living within sustainable, ecological limits. Such an arrangement is only possible when people are connected, both materially and spiritually, to the place where they live. This is the core of Thoreau's message. It is the spirit of Walden Pond.

Many people enter the environmental field because they feel a sense of loss when development ruins a childhood place. Nothing galvanizes or polarizes a community more than when a proposed development alters a symbolic place or when people discover the presence of toxic wastes. Ecological identity refers to the feelings and relationships people develop with landscapes and how they identify with nature in the process. Bioregionalism uses this relationship as the means for building cultural, political, and economic arrangements.

Gary Snyder points out in *The Practice of the Wild* that it is not enough just to want to live in harmony with nature, that to cultivate a relationship with the natural world requires awareness of the ecological nuances of place.

> The presence of this tree signifies a rainfall and a temperature range and will indicate what your agriculture might be, how steep the pitch of your roof, what raincoats you'd need. You don't have to know such details to get by in the modern cities of Portland or Bellingham. But if you do know what is taught by plants and weather, you are in on the gossip, and can truly feel more at home. The sum of a field's forces becomes what we call very loosely the 'spirit of a place.' To know the spirit of a place is to realize that you are a part of a part and that the whole is made of parts, each of which is whole. You start with the part you are whole in.[39]

Hence bioregionalism represents a powerful and practical approach to ecological identity. By discovering the place where you live, seeing the patterns in the landscape, understanding how the land changes and how humans live on the land, recognizing that humans share the land with other species, conceptualizing an ecological neighborhood—these learning activities directly expand your circles of identification. These themes pervade the literature of bioregionalism. Gary Snyder, Annie Dillard, Wendell Berry, Mary Oliver, and Barry Lopez, among others, have created a literary landscape that examines the full dimensions of living in a place. Their work is powerful because it lends poetic and literary vision to what so many people feel intuitively.

One of my students used bioregionalism as the defining metaphor for his entire tree. The leaves were a collage of writers, thinkers, and organizations that constitute bioregional thought. He has chosen a career of regional planning because it is a vehicle for lending a bioregional perspective to resource management. In describing his tree, he eloquently conveys the bioregional vision of his ecological identity.

> Ecological identity is about walking, noticing, eating wild blueberries, watching hawks, dangling tired feet in a cool mountain stream. It is about digging in dirt and getting small patches of earth "to say beans." It is mystical vision and turning over the compost. It is breathing. It is also perhaps knowing how to hunt, catch a fish, or make syrup from maple sap. I suspect it also means

knowing the limits of "civilization" and appreciating the wisdom of the hunting, gathering, horticultural societies that have constituted 99% of human time on this planet. It is certainly about knowing how economics, politics, and social conflicts shape the land and how the land, in turn, shapes our lives. Ecological identity is about not forgetting what we care passionately about in the very midst of struggling to protect it. It is, ultimately, about finding, making, and defending a home.

Although bioregionalism is a utopian vision, it serves to ground environmental thought in the real circumstances of everyday life. It demands that people connect their livelihood to their sense of place. There have been numerous practical experiments in bioregional living. In many ways this is the most interesting story of the last two decades of environmentalism. Whether it be an experiment in community technology, sustainable agriculture, urban forestry, permaculture, or restoration ecology, or the simple, practical idea of home recycling and community gardens, these practices represent hands-on environmentalism, the everyday dirty work that allows communities to collaborate, to experiment, to develop a collective ecological identity. Beyond these lofty ideas and visions there are real people at work, earning a living, raising families, trying to live ecologically sound lives, trying to contribute to community life. The various branches of environmentalism are only relevant as far as they speak to the everyday concerns of ordinary life. Environmentalism, in many respects, is a response to these basic concerns.

This is where bioregionalism merges with the social justice wing of American environmentalism. There has been a growing recognition over the last several decades of the connection between public health and the ecological health of the natural and built environments—the places where people live, work, and play. This is where people gain experiential knowledge about sustainable living, subsistence hunting and fishing, gardening, the role of recreation, the need for proper domestic and workplace ventilation and light, safe water sources, appropriate technology, and proper sanitation and trash disposal. This is where they learn that there are important power relationships that often determine the quality of community life.

These are fundamental environmental issues, and you discover your place in the community when you attend to these basic concerns. Ecological identity emerges not only through your identification with nature but with your understanding of community health and well-being. The task for the various branches of environmentalism, regardless of their underlying philosophy, is to demonstrate how these concerns reflect political, ecological, and economic dynamics. Ultimately, your actions connect you to nature as well as to human power relationships. That is the deepest measure of a person's work, the most critical agenda of contemporary environmentalism. That is the learning experience in which so many environmentalists are engaged, and it is the driving force that integrates the American environmental tradition.

The Trees Are Filled with Flowers

The environmental trees represent collectively a forest. Ecofeminism, deep ecology, and bioregionalism are merely examples of how new ideas in environmentalism are evolving. In their attempts to reconstruct perspectives on nature, community, and personal identity, they represent challenging and controversial approaches. But there are many kinds of trees in this forest. Often, radical approaches set up false dichotomies to substantiate their own positions. Deep ecology critiques shallow ecology, assuming that soft reforms reiterate old paradigms. Yet much can be learned from the conservationist tradition of American environmentalism. Concepts such as stewardship and sustainability have great depth, with much room for diverse interpretations, and many paths to ecological identity.

When I consider the hundreds of environmental trees I have seen, I am impressed at the diversity of the forest and the complexity of each tree. Most of my students eschew labels—they prefer to imagine themselves as multifaceted trees, with hundreds of leaves, each representing a different approach to knowledge, each the key to a rich and provocative learning experience. They prefer to transcend the conventional taxonomies of environmentalism and carve their own paths. Of course there are particular themes that orient people—ideas and concepts that reverberate through their life choices. But they recognize that their tree is most likely to survive in a diverse forest, with a variety of species and a rich and deep soil.

The environmental trees are blooming now. This is the colorful and fragrant springtime of environmentalism. Ideas emerge with each blossom. Over the next several decades, these trees will change their shapes, grow new branches, and display new leaves. The language and metaphors of environmentalism will continue to change.

Ecological identities will dynamically evolve. The convergence of many disciplines, the widespread acknowledgment of global environmental change, the inevitable integration of global communication—these trends will catalyze the intellectual growth of environmentalism. New words and ideas will appear. There will be countless new leaves on the trees.

And as the winds of social and political change increase in velocity, the roots will require strength and vitality. For the forest is endangered and the trees need protection. A tree cannot stand alone, and the forest is merely an island on a large planet. The expansion and integration of global economic and political systems will continue to place enormous pressure on the world's fragile ecosystems. Environmentalism will be challenged in profound ways. Its public voice will have to be strong and clear. The lessons of the environmental archetypes will require reiteration and reinterpretation. With each new leaf, the challenges and tensions of environmental work take on new dimensions.

In chapter 1, I suggested several educational paths for ecological identity work: childhood memories of special places, the perception of disturbed places, and the contemplation of wild places. Chapter 2 elaborated on those themes and showed how ecological identity evolves through the environmental tree, adding new layers as it emerges in the different environmental groups. Throughout the chapter I have indicated that the search for ecological identity leads to the path of citizenship and the domain of the commons. This represents the public expression of ecological identity—the hard work of political cooperation and conflict, the place where environmentalism meets democracy. In the next two chapters, I explore some of the places through which that path travels.

3 Ecological Identity and the Commons

Some of my most challenging and uncomfortable moments in teaching occur when someone asks me to clarify a word that I take for granted. This is particularly true of words in common usage. One part of me is impatient, wanting to move forward with the larger concept and direction. I feel slightly annoyed that I may have to slow down for a painstaking clarification. More often than not, though, I realize that by engaging the class in considering their perception of the word, an interesting learning experience will follow.

Commons is a word often found in the environmentalist's lexicon. From a teaching perspective, I find it an extremely useful word for making an important conceptual transition—how to lend a political orientation to discussions of ecological identity. In chapters 1 and 2, I frequently referred to notions such as "wider circles of identification" or "expanding the sense of self" in discussing the educational implications of ecological identity work. These concepts also have profound political implications. What are the boundaries of the wider circle and at what point does identification lead to responsibility?

I ask my class to consider the word *commons* etymologically and ecologically. Etymologically, and most simply, it refers to a basic concept—something belonging to or shared equally by two or more. What is "common" is what a group of people share together. But what is it that they share and how do they share it? Common also sprouts alternative definitions: communal, prevalent, ordinary, park; and it refers to a series of associated concepts: community, mutual, collective, public, joint. Many of these words imply a political process. People must agree as to what *is* common, and they must develop a process of assigning responsibility to managing it,whatever it, may be.

Principles of ecology demonstrate that the commons comprises an ecosystem—a complex and diverse network of biogeochemical relationships. When I breathe air or drink water, I am using what economists refer to as a common property resource. It is difficult to conceive that any single action I take, such as flushing the toilet, will threaten the quality of the resource. When I go backpacking, and camp by a pristine stream, it seems absurd that my dinner waste would negatively impact the stream. But I know that someone else may be camping downstream, or that the cumulative effect of many campers may strain the water quality. This seems like the most obvious, "common" sense. Yet the tragedy of the commons occurs because people take independent actions regarding resources such as air and water. These actions may be benign in and of themselves, but eventually the collective impact of many individuals may deplete the resource. Polluting a stream, fouling the air, eroding soil, draining an aquifer, depleting a fishery—these are all examples of the tragedy of the commons.

The commons may be a place, a resource, or even an idea (language, for example). It is such a tantalizing notion because its proximity may be ambiguous. People take the air they breathe for granted until they hear warnings about air quality. They assume that the water they use is pure and nourishing until they learn that the city water supply is unsafe. Yet it is hard to take personal responsibility for these problems, simply because air and water are so "commonplace." Unless people perceive the interconnectedness of all of their "common" actions, it is difficult to comprehend what impact their actions may have. In a sense, the commons is both everywhere and nowhere. It is ubiquitous and invisible. The commons is what people conceive it to be.[1]

The path to ecological citizenship entails promoting awareness regarding four interconnected aspects of the commons: that it is ecological (the ecosystem *is* the commons), political (it is subject to collective decisions and personal actions), psychological (one must perceive and comprehend its existence), and ultimately ethical (one must take responsibility for it). My educational strategy is to cultivate a learning community that allows my students to consider the commons from these four perspectives, based on their life experience and ecological identity work. Here is the challenge: to investigate one's perception of the commons by considering the wider consequences of individual actions and analyzing how decisions regarding the commons influences a person's life choices.

This has been the environmentalist agenda for more than a century: to make the commons the province of collective decision making and to show that private actions have an ecological impact. Implicitly, this challenges people to reconceive their most cherished notions of private ownership and individualism. Environmentalism calls attention to the various threats to the commons and demands that people take responsibility for their actions. If ecological identity work involves widening the circles of identification, then inevitably it challenges people to look at concepts such as land, community, and property from an ecological perspective. But the prevailing political culture deals with these notions from a privatized, market-oriented perspective. People have habits and perceptions about the commons that are not overcome very easily.

In this chapter I explore the idea of the commons, and its corollary, community, through the lens of ecological identity. To do so, I set up two dynamic tensions—property and the commons, individualism and community. These tensions are at the core of ecological citizenship: how to differentiate private ownership from collective responsibility, how to distinguish between what is good for the individual and what serves the larger community. If ecological identity work entails expanding the sense of self, then it must involve formulating a new context for private property and individual needs.

In proceeding through the chapter, I follow a sequence derived from my teaching experience, starting with reflective, introspective activities that allow people to explore complex and controversial ideas by virtue of their personal experience, then expanding the material to include broader theoretical concerns. The first section considers the relationship between private property and the commons. I describe a learning activity in which students compile property lists as the focus of extensive interpretation, revealing some of the psychological, ecological, and political implications of property ownership. Then I explore the dual nature of property as a concept that differentiates ego boundaries from ecosystems, and is at once both sacred and profane. The second section considers the relationship between the individual and the community. Starting with a discussion of another learning activity, the community network map, I look at how people identify themselves in a community, some of the dilemmas of community participation, and the prospects for constructing an ecologically based community matrix. Finally, the third section suggests that the formation of ecological identity inspires a politics of place, an approach to community based on a shared ecological and moral conception of the commons. These are some of the places that a collective interpretation of a complex word like "commons" may bring us to.

The Dual Nature of Property

Property and Identity: A Catalog of Personal Property

Perhaps the most direct way to reflect on the meaning of property is to construct a list of everything you own. And I mean *everything*. I ask my students to spend a few hours taking an inventory of their possessions, suggesting that they make the list as detailed as possible. Through the process of organizing this list, they learn a great deal about themselves, for they are compelled to focus exclusively on how their identity is reflected in their possessions. Of course, this is more than a narcissistic activity, or a review of one's financial worth. It is a long look at the implications of property ownership. Our possessions have symbolic meaning and lend insight to many aspects of personal identity.

I use this approach to introduce the relationship between ecological identity and the commons. Most people identify in one way or another with the things that they own and it is instructive to consider the implications of such identification. So I ask my students to write a short interpretive essay, appended to their property lists, addressing three dimensions: the psychological, ecological, and political. Some write about their feelings in compiling the list, citing how, on the one hand, their possessions provide them with self-esteem, comfort, and security. Yet there is also an element of guilt, a sense that there is something wrong about an environmentalist owning too many things. Others consider the ecological impact of their possessions, wondering what resources were used in manufacturing their stuff, how much pollution may have been involved, or just recognizing how little they know about where the things they own come from. The political dimension entails a third interpretive layer: In what way does property ownership convey power? By what process does a person come to own something, and to whose exclusion or exploitation?

When I first designed this activity, I had no idea what might emerge. Would people be willing to reveal themselves through their property lists? Or would they consider such a list too private? I assumed that environmentalists would be disposed to take on this challenge because the issue of material simplicity is such an important theme in the construction of ecological identity. Thus people should have nothing to hide. If they are willing to critique the culture of material affluence and rapid economic growth, then surely they should be ready to encounter their own ideas and feelings about material wealth. My students try to act as environmentally responsible citizens. So if they own land they take their stewardship seriously. They purchase commodities with an eye to longevity, craft, and recycling. Yet they also recognize that their property is a privilege (this often causes guilt) and they recognize the ways they are psychologically attached to it.[2]

Virtually everyone who compiles a list is surprised and even somewhat embarrassed at the extent of his or her possessions. The list itself becomes an intimidating document, as people total up the number of articles of clothing, kitchen utensils, compact disks, and other items that make up their inventory. Even people with relatively low incomes who consider themselves living at the economic margin have many more things than they thought they had. Especially, environmentalists often have a surplus of outdoor gear!

These property lists are the source of considerable ambivalence for my students. Marsha, who works for a recycling consulting firm, cites her contradictory emotions:

> On the one hand I feel overwhelmed, burdened by the necessity of taking care of all these possessions, worried about their potential loss, theft, or destruction. Add to this the fact that it doesn't lessen my burden to throw anything away. I feel guilty about the ecological repercussions of throwing reusable, potentially repairable items into the landfill. On the other hand, my possessions also make me feel prepared. My car can get me where I need to go without depending on anyone else (save a fuel supplier). My variety of clothing allows me to dress for the randomness of the New England elements. My computer helps me meet the writing requirements of my graduate work. My books allow me to reach into other minds and other worlds. Some of my possessions also give me a sense of comfort—my futon, my "fuzzy" blanket, some very soft and comfortable clothes, my tapes and CDs and stereo equipment.

Marsha understands the extent to which her property is a symbolic representation of her identity. Her clothes enable her to cultivate multiple personas (working professional, student, daughter), her books and CDs reflect her cultural tastes and interests, her economic mobility reflects her class. She is reflective enough to recognize how much she enjoys her possessions, yet she also acknowledges their ephemerality and superficiality.

Carol, a high school biology teacher, realizes how she uses her property as a means of power, both to attach significance to her education and upper–middle class upbringing, and to achieve the perception of what she calls "independence."

> I have property which gives me entrance into the world of the elite. Great-grandmother's fine china, Uncle Nash's sterling silver, my money, and the crystal bowl from Tiffany's contribute to my sense of superiority over the average Joe. It doesn't matter who Joe is. If he calls me a bitch, I can dismiss him and his comments. He is low-brow, ignorant of the ways of the upper–middle class, and so not deserving of my attentions or my concerns. I have used my property to insulate myself from him. At the heart of owning property is the independence it gives me. My car offers me independence in transportation. My IRA provides me independence from a failing Social Security system. My VCR affords me independence from the local cinema. My journals give me independence from relationships with other people. America holds independence as its Holy Grail, that which is most desirable but ever elusive. We pursue property not only for the pleasure it brings us but for the power it buys us. The more power we have, the more independent we are, and the closer we are to fulfilling our quest.

Carol acknowledges the political tensions inherent in property ownership. Private property is perceived as a guarantee of security, a symbol of autonomous activity, and a safeguard of liberty, yet it is also the vanguard of imperial power, elitism, and an expression of unbridled affluence. Her relative wealth is a privilege and a trap, a means of independence and a tool for exclusion and insulation.

As the students interpret their property lists, the most common theme is the tension between property ownership and "politically correct" environmentalism. Most people have difficulty reconciling the fact that their materials come from the earth which they are trying to protect. Surely, this is abstract and idealized, yet the connection is inevitable. Material goods involve the extraction of resources from the ecosystem. It is entirely conceivable that the production process which yielded the good contributed to environmental pollution.

Pam, a graphic artist, cites the hypocrisy of her environmentalism:

> Life as an environmentalist in the United States seems to be a balance of intense, narrow, and focused awareness with broad, blissful ignorance. We attempt to act in ways that express our understanding of as many human rights, environmental, political, bioregional, and health issues as possible, yet we fail to see the implications of our resource consumptive lives. Even with all my efforts to reduce my environmental impact, I rarely think about the amounts of trash produced, resources transformed, energy consumed, and pollution emitted to make my life as convenient, rich, and comfortable as it is.

Pam understands that her wealth is derived from the ecological commons. If she is to live a morally consistent life, she must reevaluate the economic and political conditions that contribute to her wealth. At whose expense is it derived?

Carol makes a similar point, describing the interconnections between her property and the global ecosystem:

> Everything I own was made from material pulled out of the ground. Plastics made from petroleum, metals made from ores, fabrics made from plants deriving nutrients from the soil— all are products which once belonged to the earth. Through nasty chemical processes and transactions involving copper, nickel, silver, and trees these objects become mine. The kitchen gleams with stainless steel, a metal that was once iron ore. The local ecosystem was destroyed to mine that ore from the ground. Another ecosystem was destroyed to mine the coal from the ground, and the coal was then burned to process the iron into steel. More energy was used to transport the steel to Belgium, or Japan, or Taiwan where the laborers were paid dirt wages to hammer out appliances for Americans to buy. The finished products were shipped back to the United States. The total energy input (fossil fuels) into these things of mine is huge.

The point of this property list activity is not to coordinate a collective orgy of environmental guilt. Rather it is to reflect deeply on the inner meaning of property as a tool for connecting ecological identity with perceptions of the commons. Most people use their possessions as a means to construct personal identity. But the search for ecological identity challenges that narrow interpretation. Every object a person owns may have utilitarian and symbolic value, but it also has ecosystem value. It was fabricated from resources derived from the commons. This is another example of what it means to widen one's circle of identification. For example, a "personal" computer is not just a reflection of one's persona, a tool for writing, a menu of interests and personality, or a vehicle for global communication. It also represents the complicated international production process and all of the costs and benefits that go along with it. Personal property is inextricably tied to the commons.

Yet the idea of ownership may preclude this broader identification. Often people confuse *ownership* with *control*. Alex, who has a career in sustainable agriculture, realizes that he has to reflect carefully on this dynamic, as it prevents him from building a more sustainable relationship with his land. This realization liberates him to identify with the land rather than his ownership of it.

Material possessions are supposed to make us feel safe, a part of something, or well prepared for something. They're supposed to keep the monsters at bay, the wolf from the door. Our possessions are the door. I am consumed by my possessions, even my land. I take care of my land responsibly but in my own way, not the land's way. It has become something I've created out of an idea. I have transformed it to serve my own visions. I subvert it. I know this. I am aware of this constantly. Even in my sleep. What is my relationship to this land I own? I'm beginning to get an inkling of what it is. The land is not mine. It indulges me. It seeks me out as it seems to bend to my work at it. For 7 years I have watched the subtle changes in this land. I have seen it approach me in a number of ways. This land is an opening and it will guide me if I let it.

To own something is to create a distinction, separating what belongs to you from what belongs to someone else. In Alex's case, this connotes the presumption of control. Because he owns the land, he assumes he can use it any way he wishes, reflecting his vision and ideas about sustainable agriculture, farming the land in his own image. But Alex is also developing his ecological identity. He has a broad interpretation of the ecological commons. He knows that ecosystems have no boundaries, and the distinctions that he has made serve his ego and self-esteem.

Alex, Pam, Carol, Marsha, and the rest of my students struggle to reconcile this profound question of personal identity: how material possessions convey a sense of individual worth and facilitate the expression of individuality, affording them self-esteem, security, and a degree of personal power (including using their things to promote their ecological values), yet also create distinctions which allow them to deny or ignore the more exploitive political and ecological consequences of ownership, forgetting their responsibility to the ecological commons and the broader circles of community.[3] They are striving to bridge the conceptual gap from ego boundary to ecosystem, trying to forge paths of personal development that coordinate their own needs with their responsibility to the ecological commons. And as I scan their property lists and review their essays, I realize the extent to which their struggles reflect my own.

From Ego Boundary to Ecosystem

I live in a small house on 5 acres of land in the rural hill country of southwest New Hampshire. I consider this "my" land because I bought it (with the help of the bank); I dwell on it (with my family); I have "improved" it (with a garden and landscaping); and it provides me with a measure of safety, independence, and individuality. The land is shaped like a wedge of pie. One side adjoins a dirt road, providing ample space between neighbors. On another side, I am protected from development by a large beaver pond, an exquisite open wetland that is devoid of human settlement. Behind my house, facing east, the land descends a small hill and eventually ends at a point at which five different property lines converge. The terrain is thickly wooded with second-generation tree growth, interspersed with numerous stone walls. You could categorize the land as "hardscrabble" or rocky, glaciated soil, too difficult to farm. The land was previously cleared and used for grazing and timber, but like many farms in the hill country of northern New England, farming was not a viable commercial venture. I have come to know this terrain intimately, observing its natural history, its rhythms and changes, its coinhabitants, and the microclimatic nuances. Unquestionably, I identify with this land. It is part and parcel of my ecological identity.

As I sit in my study, I am surrounded by another kind of territory, a room that I have carved out as my own. This is the place where I write and read, where I work on my computer, where I store my books, games, and assorted collections. It is intrinsic to my personal identity. I "own" all of the things in this room. Whether it's the artwork on the wall, the way the bookshelves are organized, what I choose to keep and what I choose to move along, these artifacts are powerful reflections of how I perceive myself, a reminder of how much of my identity is tied up in my possessions.

There is a small picture window which allows me to look out into the dense northern hardwood forest. So I alternate my gaze, moving from the bookshelves to the computer to the large oaks and maples, inside and outside, within and without. Where does my property end and nature begin? How do these distinctions and boundaries allow me to define myself? I am gazing at my "property," understanding how, on the one hand, I have been deeply socialized to the idea of individual ownership, how it enables me to symbolically differentiate myself, defines my individuality, lets me make value statements about who I am and what I believe in. Then I make the intellectual switch. I think about this land as it has changed through eons of geological time. I ponder its recent history and consider its relationship to a larger ecosystem.

The idea of property is a convenient illusion. It represents a distinction, a differentiation, an ego boundary that allows me to separate what is mine from what belongs to others. Yet I know that "my" property is the essence of impermanence. My land and belongings are projections of personal identity, cultural configurations that bring a sense of order to my life. I know that I own this land, this house, these possessions, and I am grateful for each of these things. But I do not derive pleasure from the idea of ownership for its own sake. Rather it is the process of learning about my land and enjoying my home that makes these possessions meaningful.

How is it that I was able to buy this land? I was fortunate to grow up in a middle-class home. My father was a dentist. He put metal (mined in some remote place) in teeth (so people could eat their food without pain). People paid him for that service. Where did they get their money from? Undoubtedly, some of those people got their money from creating some form of ecological distress. They paid that money to my father who was able to send me to college where I received intellectual training. Now I work at a private university that is mainly supported by tuition dollars. That money supports my salary. Where does that money come from?

Clearly, it's absurd for me to think of this as "my" land, yet I am bound by this odd duality. Ecological identity work helps me understand the meaning of this land as it places the concept of ownership in ecological perspective. As long as the land is in my "stewardship," I can respect it, understand it, learn about it, and use it sustainably—approaching it from this perspective whether I use it for subsistence or as a bucolic retreat. Nevertheless, I am always aware that this land exists within specific boundaries, that I can sell it if I choose, and that it has a monetary value. And I am attached to this ownership notion. I can accept limitations on my wealth, various forms of taxation for the public good, even some restrictions on how I can use my property, but I can also imagine myself feeling stripped bare without the security of personal ownership. I would like to think that I can live for pure ideals and that my values about nature would transcend this attachment to ownership. But the idea of the commons remains intangible and abstract. In many respects I would be lost without my land or possessions.

And so would many others. Property represents an important form of security against the potential tyranny of the state. There is historical precedent for this. With the growth of industrial capitalism and the decline of monarchy, the possession of property became fundamental to liberal democracy. People could accumulate wealth, experience autonomy, and derive prestige through their acquisition of property. Fred Weinstein, in *History and Theory After the Fall*, explains how the possession of property contributed to the rise of the bourgeoisie, and how their ties to property transcended power and income.

> Property served in both a directive and protective way to establish and confirm their sense of what was good, worthy and admirable about themselves and their class. The bourgeoisie have therefore not feared the loss of property per se so much as they have feared an assault on their integrity and esteem, and this is why they are so emotionally affected when their rights to property are challenged or threatened and why they so readily express anger, alarm, anxiety and frustration, and so on in the face of social change; this is also why they so typically claim and feel that they are defending a moral position rather than an interested stake in property as such.[4]

These observations are intended to demonstrate how difficult it is for most people to give up what they own, or the prospects of doing so. The *idea* of property, whether it refer to land, money (a consolidated form of property), or personal belongings, is fundamental to self-esteem. That is why people feel violated when they are the victims of theft. It is a challenging conceptual leap to move from what is "mine" (the boundaries of personal property) to what is "ours" (collective ownership of the commons). People think that they can best determine what to do with *their* things and they are hesitant to relinquish property to any collectivity, especially the abstract state. Hence Americans are suspicious of taxes. Without a collective notion of a common good, Americans will only grudgingly and painfully contribute to the state, the town, or any tax-collecting jurisdiction.

Yet whatever an individual, state, or corporation owns must in some way be linked to an ecological, historical, and political process. Property is nothing more than a claim on a resource. Homes, cars, clothing, and all forms of material wealth, including so-called intellectual property (books, software, designs) involve the transformation of natural resources. The claim is a proclamation of ownership. With this claim come certain rights, what we usually call property rights, regarding how the resource can be used. If the use of this resource affects other people, it is likely to have more regulations and responsibilities attached to it. Some resources are so "common" that their projected use becomes the source of controversy.[5]

These controversies are critical to environmental politics: the boundaries of ownership, decisions about common property resources, the juxtaposition of the public good and private property rights. The environmentalist calls attention to the ecological ramifications of decisions about property. In many cases, this portends policy suggestions or legislation that involves the public regulation of private property. For example, a business might own a factory but it is subjected to regulations and restrictions regarding its effluent wastes, possibly inhibiting the short-term, economic efficiency of the industrial process. It is no wonder that there is so much literature regarding property rights and the commons.

In order to delineate a national park or a wildlife preserve, the state may create land use restrictions, or even authorize the purchase of land, against the will of the landowner. Private landowners who have strong ecological values may be more willing to accept limitations or guidelines for the use of their land. However, if a person is more profit-oriented, owning the land as an investment, or if property exclusively serves one's self-esteem, such ecological guidelines will be irrelevant. They will be perceived as an infringement on personal rights. Or, and this is quite common, if landowners are rooted to the land through generations of residence and work, but face various economic pressures, they might perceive themselves as unable to comply with ecological limits without having to sell the land. Finally, there are situations in which property is owned by a speculator or developer who may have nothing but an economic connection to the land. In many cases, ownership claims are purely profit-oriented.

Multinational corporations own large tracts of land and typically do not take kindly to environmental groups who tell them how they may or may not use that land. From their perspective, the land represents an investment in natural resources, and they are the ones to determine how it is best managed. Conceivably, they will use "ecologically sound" resource management techniques, but might object in principle to any state restrictions or guidelines. Typically, however, natural resources are perceived as economic investments—not integrated ecosystems—and a clash between economic and environmental interests emerges.

This emphasis on profit and investment does not only reflect the myopic profit orientation of unbridled capital. It affects small landowners, businesses, communities, and just about any individual or group that is involved in economic activity, by virtue of identifying their land as a commodity, as a source of profit. This fuels the so-called wise-use movement, which is largely funded by big business but comprised of small landholders. They are troubled that anyone would consider restricting the use of their land. The rage and anger which pervades these groups indicate that they feel, in some way, that more than their economic interests are at stake—their freedom, autonomy, self-esteem, and self-respect are threatened. They consider it intolerable that someone should challenge their right to take care of or use their land, or restrict their economic opportunities. Environmentalists receive the full vent of their fury because they are seen as the political actors who most dramatically critique the use of the commons. It is environmentalists who wish to preserve land for the public good. They are viewed as intellectual elitists who think they can tell other people what is best for them. In this way Greens become Reds, as any talk of making the private become common sounds very much like Communism.

To propose the regulation of land is to imply a restriction of power.[6] But often, the unrestricted use of land leads to the misappropriation or exploitive use of natural resources. This is the crucial problem of environmental politics: developing approaches to the commons that maintain the sanctity of property rights, thus protecting the individual, yet also recognizing the integrity of the ecosystem. Property may serve a useful function as an appropriate expression of limited territoriality. But how can that expression be accomplished without sacrificing the ecosystem and without falling prey to the accumulation of property for its own sake?

The idea of property is filled with contradictions. We are simultaneously liberated and imprisoned by property. It protects us from the tyranny of the state just as it is used as a form of coercion. It is a means of personal wealth just as it involves the prospect of exploitation. Property is a source of pleasure when it enables people to enjoy the use of a resource; it contributes to suffering when pleasure is derived from its exclusive use.

As a child, I remember witnessing fights between neighborhood suburban children. In the end, someone would always shout "get off my property." How could a child make such a pronouncement? In the absence of physical strength, this is all the child could resort to. It was his final and most desperate option, his last chance to reclaim his self-esteem. But because the child lacked legitimate authority and could only enforce his claim with brute strength, he would be ridiculed (at best), and clobbered (at worst). He could attempt to draw the line where the lawns met. This was a desperate attempt to convey power through the boundaries of the property marker.

The accumulation of property conveys the illusion of power. Men and women secure the bounty of nature through their ability to appropriate its wealth for themselves. Benjamin Barber in *Strong Democracy* describes how the psychology of property and the functioning of liberal democracy are deeply embedded in the proprietary claim.

> Power-seeking, like freedom-seeking, finds its natural extension and logical end in property acquisition. For property is a form of cumulative power, an authoritative variety of institutionalized aggression, by which the claims of individuals to adequate means are given a permanent and legitimate home. Raw power is time bound; it yields only temporary control, providing a transient security that lasts only as long as the direct mastery of animal over animal can be sustained. Possession lends to coercion a security over time, enhancing the effect of power by diminishing the need for its perpetual exercise. What we merely "need" from the common environment, we must beg, borrow or take—over and over again and at continual risk. What "belongs" to us, we only have to protect and use, and at much less risk.[7]

The fundamental unit of property is the home territory. Our homes are in places. And these places have an ecological setting, the habitat. In ecological terms, habitat refers to the native locality of a plant or animal, including its community context (feeding habits, range, proximity to predators, terrain, etc.). The word is derived from habit, which refers to one's mental constitution or settled disposition. From habit we also derive inhabit (to dwell) and habitation (dwelling).

The opposite of home is homelessness. This is different from a nomadic existence in which a group of people moves from place to place according to the season. To be homeless is to lack a place. It is the involuntary severance of one's connection with a home territory. Homelessness is the shadow of property rights. Once a person claims ownership of a piece of land, other people who live on the land can only live there according to the terms of the owner. In many cases this is an intolerable situation. Urban homelessness, in particular, is a chronic problem of liberal democracy. Similarly, nothing can be more humiliating than a home foreclosure. These situations are not merely trends of the late twentieth century. Industrial capitalism was made possible in part by the enclosure system, which evicted thousands of tenant farmers and forced them either into the mills, emigration, or homelessness. The exploitation of the tropical rain forest has transformed indigenous peoples, who dwelled in habitats for generations, into urban homeless, flooding the cities with people who have nowhere else to go.

So, on the one hand, property allows people to be rooted, to have a home, to distinguish their place as a home territory, conveying independence and safety. On the other hand, the system of property rights leaves many people homeless, uprooted, and enclosed. If we are fortunate enough to own property, we may enjoy that property, but there will always be a measure of guilt if we contemplate the full political and ecological implications of our ownership. Is it possible to reconcile the dualities of this complex web of property?

Property Is Both Sacred and Profane

The lens of ecological identity can be used to integrate the dual nature of property. By compiling property lists, my students realize how *important* their property is in contributing to their individuality. If we consider property from the standpoint of ecological identity, then we understand how what we own contributes to an expanded sense of self. A person's property lines are merely a land use boundary. Why shouldn't a person identify with both the land *and* with ownership of the land? By integrating both perspectives, one can derive a sense of political responsibility to the ecological commons. The land is not here just for personal satisfaction and benefit; it represents the ecological relationships and human interactions of the entire biosphere. By thinking carefully about where your possessions come from, and the ecological circumstances of their appropriation, you connect what you own to a larger sphere of relationships. In this sense, property becomes a vehicle that links the ecosystem to the individual.

What happens if we take our property really seriously, as a means of constructing personal identity through increasingly wider circles of identification, in this case, the perception of the ecological commons? Jacob Needleman has written an extraordinary book called *Money and the Meaning of Life* in which he asks a similar question regarding money. If we view money as a consolidated form of property, then his observations become instructive. Needleman describes the human condition as inescapably bound by two conflicting demands, the outer world of social and economic exchange—what we usually refer to as the material world—and the inner world of love, knowledge, and creativity, or what we usually refer to as the spiritual world. He explains that human life is most meaningful only when we occupy both worlds at the same time. Money is the currency of one of these worlds, hence its profound significance. Yet this force alone cannot bring meaning into human life. "Meaning," Needleman writes, "appears only in the place between the worlds."[8]

According to Needleman, money was invented as an instrument for linking material life to the life of the entire community, to facilitate interdependence. "It was meant only for helping people directly to live in the material world, while at the same time recognizing their dependence, first upon God and then upon each other."[9] This is why, Needleman explains, the earliest coins bore both a religious and secular symbol. But modern coins no longer have a sacred side; money has become a purely secular force. Thus money is cut off from any spiritual aspiration; it no longer serves to bridge the two worlds.

Perhaps by looking at property in the same way that Needleman looks at money, we can make the leap from ego boundary to ecosystem. Property is both secular and spiritual. A landowner may earn a living by working the land, but that land also connects him or her to the diversity of life—a reminder of the interconnectedness of the ecological commons. Just as property serves to reinforce autonomy, it continually reiterates interdependence. It grants a degree of power, but only within the ecological limits of the commons. People can identify with their property, but that identification must be broad enough to include the polis and the ecosystem. It is this deep awareness of property that balances the individual and the collective: the need to earn a living, with reliance on a complex ecosystem; the recognition that property ownership is a profound responsibility.

Consider the property/commons feedback network (see figure 3.1). I refer to this as a network to show the interconnectedness between the various elements in the diagram. On one side of the network is the commons, which represents both the ecosystem and the various information pathways that allow humans to learn and communicate. From the information pathways people derive knowledge, which is essential for their ability to understand the ecosystem, and also provides the skills to pursue an occupation. From the ecosystem comes the land and natural resources that provide sustenance. On the right-hand side of the network is self-esteem, or the personal integrity that is the right

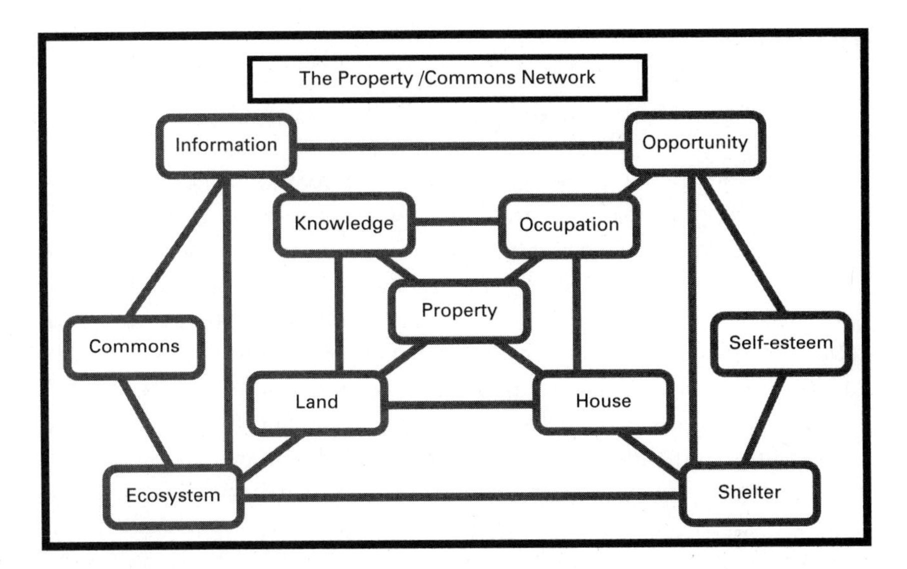

Figure 3.1
The property/commons network

of every person. People require shelter, which includes food and basic material needs, and opportunity, which represents the education and the choices which allow access to homes and occupations. The diagram is designed to show how these elements are interconnected. Ultimately one's self-esteem depends on one's share of the commons.

Property is in the center of the diagram because it is the most crucial element. On the one hand, it allows people to make the distinction and differentiations that symbolize individuality. Simultaneously, we see how the notion of property is intrinsically connected to all aspects of our existence. The idea of a feedback network is derived from ecological thought and information theory, conveying, in this case, the interconnectedness of property, how all aspects of property ownership and individuality affect the ecosystem and the commons. This diagram represents a way to view property through the lens of ecological identity, displaying the loops and circles of broader identification. This is the basis of a political awareness that links ego boundaries to ecosystems, that allows for individual autonomy and collective responsibility, that connects personal action to ecological process, and that allows property to be at once both sacred and profane.

The Fabric of Community

The Community Network Map

The commons also represents a network of relationships, or what is typically referred to as community—the people and species with whom we interact on a regular basis, or with whom we share the place where we live. *Community* is another word in common usage with broad connotations, requiring considerable clarification and interpretation. One's perception of community is crucial in traveling the path of ecological citizenship. This inspires several layers of reflective inquiry: a discussion of the relationships and networks that constitute a community, considering the dilemmas and challenges of community participation, and developing the criteria for a cohesive community. This section explores the changing context of community in postmodern life and how ecological identity work can lend greater definition to community responsibility and participation.

I find that the most effective way to interpret and discuss community is to ask students to draw a map or chart depicting their community relationships. I encourage them to consider all the possible matrices of their perceived communities: affiliations and associations, places, people, neighborhoods, habitats, electronic networks, whatever they consider to be a representative portrait. I urge them to use their imagination and draw attractive maps, using colors, charts, and arrows to highlight the various relationships. These maps reveal an extraordinary range of personal expression, including three-dimensional blocks, templates and overlays, Venn diagrams, pinwheels and mobiles, as well as more typical flow charts and cluster diagrams.

This is a surprisingly complex task. Many of my students (similar to most Americans) are highly mobile, changing residence and location frequently, finding it difficult to establish permanent relationships in a given area. There are reasons for this—job availability and educational opportunities—but it runs counter to their professed long-term goals of developing ecological intimacy and lasting interpersonal relationships in a specific place. Second, many people experience a conglomeration of far-flung networks and associations, based on interests, hobbies, or friendships, that are linked through advanced forms of electronic communication. So their community network maps reveal a constellation of personal orbits, each representing a significant aspect of their life experience.

To interpret the maps and to use them as a springboard for group discussion I provide a list of questions, organized around two broad themes:

The Structure of Community Networks

- What makes any network a community?
- To which communities do you feel emotional attachment?
- Who belongs to your communities, people like yourself, or people who are much different from you?
- How do the communities to which you belong interconnect?
- Do they form a tapestry of cooperation or do they underscore isolation and division?

Belonging to a Community

- Are you as active in community life as you'd like to be? Which of your community activities are political? Which involve conflict? Which involve environmental issues?
- What are the prerequisites of community membership? How do your community memberships change?
- What allows some communities to change and others to avoid it?
- In what ways does your perception of community reflect your ecological identity?

In surveying the array of maps, the first thing I notice is the complexity of the networks. Some people are firmly rooted in their town or neighborhood, taking an active role in community issues: recycling, the town meeting, school functions, local elections, and so on. Others have only the most superficial local relationships. Instead they highlight their professional affiliations, their correspondences, or their friendships, which may encompass a vast geographical area. I observe the growing importance of electronic communities, people who communicate regularly on the Internet, forming collaborations, affiliations, and friendships.[10] For some people, the workplace is a strong community, especially when the staff and leadership share a common vision based on a long-term commitment. The element of ecological place is critical for some: they perceive their most important community as the people, species, and organisms that are their neighbors.

And there is considerable ambiguity as to what constitutes a community. One student is enamored with his proliferating files of e-mail, describing the relationships he has built using electronic communications. Brad, a computer programmer, proudly describes how he has developed a computer network of environmentalists, who are both friends and colleagues, in some cases, people he has never seen in the flesh. He considers this a community of activists, people who have similar political goals and objectives, and who eventually come to know and care about ane another. He observes that he could never establish such a community within his suburban dormitory town, where he has nothing in common with the local residents, and where he feels alienated and isolated.

Susan, a community activist, argues that such a virtual community is ultimately ephemeral, no deeper than the pixels on the computer screen, consisting of a homogenized group of electronic talking heads who may have similar values, but have only superficial relationships. She asserts that a community must be based on the place where a person lives, the various neighbors and species that share the land, and the issues and experiences they have in common. Electronic communications prohibit the building of community, engendering fragmentation and hollow human contact. She says, "The most tangible community emerges as a result of a person's daily habits. Neighborhoods share a geographical space. They have a common ecological habitat. Community meetings, local schools, town dumps, markets, corner stores, block associations: these are the backbones of local communities. You can be seen and heard. You can sense body language and pheromones. None of these relationships are possible with electronic communications."

Brad argues correspondingly that, "proximity may not breed familiarity. And familiarity can often breed contempt. It is easy to indulge in romantic fantasies about delightful community life, but small towns or neighborhoods can be repressive and many people flee their communities so they may live as they please."

Susan cites the work of Robert Bellah and his colleagues, whose *Habits of the Heart* deals with the tensions of individualism and community in American culture. The authors explain that the term *community* is used very loosely by most Americans. In contrast they propose what they call a "strong definition":

> A *community* is a group of people who are socially interdependent, who participate together in discussion and decision-making, and who share certain *practices* that both define the community and are nurtured by it. Such a community

is not quickly formed. It almost always has a history and so is also a *community of memory,* defined in part by its past and its memory of the past.[11]

Brad suggests that communities such as these are merely anachronisms, relics of an older era, reflecting the demographics of a lost century and a nostalgic ideal. He argues that the twenty-first century will bring new cultural expectations to the idea of community. The potential of shared experience will be facilitated by electronic communication. People will form intimate relationships with neighbors in distant locales. The immediacy of shared place will be linked by the common perception of global environmental change, as people recognize that they all live on the same planet and ecological pollution transcends parochial boundaries.[12]

Kenneth Gergen, in *The Saturated Self: Dilemmas of Identity in Contemporary Life,* points out that many superficial networks might appear to serve as communities, whereas they are merely temporary associations of like-minded individuals.[13] Gergen categorizes these associations. He describes *collage communities* in which people temporarily live in the same neighborhood, but have nothing else in common. They move in and out, have different values, and develop little affiliation for the local place. *Cardboard communities* exist when all the "trappings of face to face interdependence are maintained, but the participating bodies are absent." For example, there are churches, shopping malls, and community centers, but the houses and apartments are largely empty. They serve merely as bedrooms for a highly mobile population. Finally he describes *symbolic communities* in which people communicate with others based on like-minded affiliations such as sports teams, electronic churches, or particular hobbies. In *Habits of the Heart* the authors refer to *lifestyle enclaves,* communities in which people have chosen to live together because they value a shared lifestyle. Yet they may have little else in common. All of these superficial, but prevalent pseudocommunities provide little context for sustained political interaction or deep personal commitment.[14]

Susan and Brad agree that there are many different ways to conceive of a community and that at some point the class should develop criteria for what they consider to be cohesive, interdependent, participatory community life. But before we can get to that, another issue emerges, the tensions that separate the individual from the community. We begin to discuss the quality of community membership and the dilemmas that inhibit more comprehensive participation. Few people are as engaged in their communities as they would like to be. Some rationalize this, making all kinds of excuses—lack of time, career, family, schooling. For others, a perceived lack of community engenders a kind of loneliness; they are concerned that something important is missing from their lives. An overriding theme emerges: the anxiety and fear that prevents people from participating in community life.

Robert, a field ecology researcher, reveals the importance he places on independence. He perceives community involvement as sacrifice.

> I value community but I have always been taught to think independently and ultimately my individualism comes before any community I might belong to. Some people are joiners. I am not, particularly if it means that by compromising my individualism, I am disempowered. Of course there are also situations where for the good of the community I would be willing to sacrifice my individualism, if I believe that is what is called for. Community life can be dangerous and deceiving if the line between community and the individual becomes blurred. There is no doubt in my mind that community life represents service and sacrifice. But can there be sacrifice without suffering? People who choose to serve and make such a sacrifice should believe passionately in their cause.

Robert is legitimately worried about the dangers of community life. Despite his deep commitment to environmentalism and his sophisticated understanding of the commons, his emphasis on individualism is clear. He understands this about himself. He wonders if threats to the ecological commons will be the impetus for him to overcome his fear and to look differently at his community responsibility. Yet he admits that in such a case he would only take action because of his love of nature and his desire to protect himself. Perhaps this is exactly what jeopardizes the commons, the failure of people to sacrifice their independence for the risks of community involvement.

Kristin, an environmental educator, who has spent the last 10 years working at seasonal jobs, realizes she has never experienced "authentic" community life. Her challenge is how to construct a community network when she has never really been taught how to do so. How can she be a total participant when for most of her life she has felt like an outsider?

> I am on the threshold of community connectedness, but I find that it is an elusive edge. As a child I yearned for a group of friends. Always and forever I seemed to watch that ideal from the outside, never quite able to find or grasp it. I am a partial participant in isolated communities spread across time and place. My ties are few, but they run deep—ties to family and a limited number of friends.

Priscilla reflects on the boundaries that she creates to separate herself from other members of her neighborhood, and how easy it is to be lazy, luxuriating in an illusion of independence and self-importance, not taking the time to become involved in community relationships.

In my particular case, I believe that it is the relative independence of all the individuals in my neighborhood that has caused them to become so isolated and alone. By being totally self-reliant, we no longer find it necessary to relate to people that we do not have to, or to assume the social burden of having to care whether or not one of our neighbors is in need of help. Since we don't need them to ensure our well-being, why should we invest our time or energy in undertaking the difficult task of having to get to know new people and interacting with them on a daily basis? It is far easier to ignore those people whom we do not need to know and get along with, and to keep whatever long-established routine of life that we have always had.

And finally, Marsha expresses how one part of her would love to become more involved, but she feels that with all of the professional demands on her time, her most important priority is to look after herself. She describes two sides of her personality, the champion of individualism and the diehard community fan.

The Champion of Individualism shares some of the Diehard Community Fan's views and is concerned about a lot of community issues: illiteracy, the writing off of our society's elderly population, abused children, battered women, pollution, overuse of natural resources, poor land use planning, and more. She thinks about volunteering in a soup kitchen, or volunteering to read to elderly citizens at a rest home in her neighborhood. She thinks about being a Big Sister, or helping adults in need to overcome their illiteracy. She wants to work with citizen groups trying to address pollution issues at a former air base and an active naval base near her home. But each day starts with a lengthy "to do" list and ends with a multitude of things undone. She is juggling too many balls and can't imagine adding another one labeled "community service" to the mix. She'd collapse.

Here is the dilemma. All of these socially and environmentally concerned people who are devoting their careers to environmental work, who understand the responsibility they have to their communities, who are striving to expand their sense of self in relationship to other people, nature, and the commons—they either cannot find the time to participate in community life or are worried about the risks of community involvement. They realize how much inertia they must overcome and how the culture of individualism pervades so many of their choices. In addition, they gain empathy and understanding for the great numbers of people who are totally intimidated by the prospect of community participation.

The problem is to transcend these dilemmas. Some of my students argue that the key to balancing individualism and community participation can be found by actually making a strong commitment to community life. This takes courage, support, and perseverance. For years, Carol has made small donations to environmental organizations, reassuring herself that this kept her involved in community issues. But at some point she became overwhelmed by the number of impersonal mailings she would receive asking her for money, and she had a feeling of emptiness, as though she were being used. She realized that this was not a substitute for community participation and she will only receive in proportion to what she is willing to give.

I am slowly realizing this: a community only gives back to an individual what that individual gives to the community. Simply writing a check to make one feel involved is a shallow action with shallow returns. The emptiness in my community network echoes the insecurity in my heart. Until I am willing to commit myself to a community, no community will commit itself to me. . . . Engaging myself in an action-oriented community requires my willingness to take action, to make room in my life to fill the emptiness.

Amy, a botanist and environmental science teacher, explains that she grew up being taught the virtues of community involvement, and she uses the role model of her family members as a reminder.

Connected with and housed in the church was South Louisville Community Ministries, an organization that helped people in need to obtain housing, utilities, food, and other services. My grandfather served as their director for a while, and my mother was also very active. I remember as a child helping her sort canned goods and clothing in stuffy dark church rooms, and I know this experience has been vital in the development of my sense of community and of service. (Incidentally, my 80-year-old grandparents still deliver meals-on-wheels to the "old people," as they say.)

Several people explain how they are working to reconceptualize their vision of community through the lens of ecological identity. They try to see the wider context of their actions, understanding how their attempt to identify with nature brings them into a wider circle of human relationships. Jennifer, an environmental educator, uses the community awareness map to create a metaphor for personal identity:

If community is defined by relationships, how do relationships define self-identity? When I look at my relationship, I see a series of concentric circles radiating from a center. The center or origin is an individual: a physical, thinking, and feeling being. But is there really a center? By definition relationships cannot occur in isolation. So the circles of my relationships overlap infinitely with others. I have diagrammed my community relationships somewhat like waves moving away from a pebble dropped in a pool of water. I am that pebble, dropped gently, a few inches from shore, into a pool whose far side is too distant to see. The pool represents all possible relationships, human and otherwise.

Doug, a furniture maker and a concerned activist, is trying to become more involved in town politics. He made a detailed map of his local town, integrating his human networks with a drawing of the watershed. As a relative newcomer to the town, his challenge was how to get involved in local environmental issues without seeming presumptuous and inappropriate. So he spent several years carefully observing his neighbors, slowly building relationships, and trying to demonstrate his commitment to environmental issues, without being doctrinaire or judgmental. He realized that the watershed became the source of his community connections:

I did not come to it quickly, but gradually I've become connected to a broad and awakening region defined by the watershed boundary. I am as connected as I've ever been through this passion for a river and I'm becoming more active politically because of it. I didn't want to be some busybody do-gooder pulling tires and junk out of the river and shaking my fist at the perceived offender. Instead I found a river at risk and full of life. I found ecological and political connections.

What emerges from these maps and dialogues is a new depth of understanding. When people are willing to explore their deepest feelings about community participation, they can face the tensions and ambiguities that pervade individualism. This is yet another layer of ecological identity work, the ability to confront the contradictions and inconsistencies of ordinary life, understanding the cultural impediments and the personal habits that constrict the circles of identification.

The community network maps set the stage for a more comprehensive discussion of the function and purpose of a community, how community participation is linked to the perception of the commons, and how the formation of ecological identity integrates environmentalism and citizenship.

An Ecological Community Matrix

What makes any constellation of relationships a cohesive community? There are many factors that might contribute to one's choice of community orbits: associations of like-minded people, narcissistic projections of the many facets of one's identity, anonymous and temporary intimates, people who discuss important issues that they have in common, people who share the same geographic region or electronic space. What distinguishes any group of individuals, either in an electronic or a face-to-face context, so as to call them an authentic community?

Cohesive communities require people who are willing to challenge and respect one another, who can develop short- and long-term commitments, who take responsibility for their shared (common) space, and who perceive that their collective actions make a difference in the quality of their lives. Community membership is not possible without the motivation to participate. People must believe that the risks of involvement, including the time spent and the exposure to public controversy, broaden the possibility of their life choices and widen their circles of involvement. Hence participation opens the possibility of expanding the sense of self to include a larger matrix of concerns.

For the environmentalist, the idea of the commons is intrinsic to community; it is a crucial concept that links ecological identity to community life, and thus serves as a context for political participation. Herman Daly and John Cobb, in *For the Common Good,* are interested in the structure and function of what they call sustainable communities. They are concerned with how to construct an ecologically based economics that is rooted in bioregional community relationships. In order to accomplish this, Daly and Cobb are interested in the personal process of how people move from "individualism to person-in-community." For them, community is a term which suggests "that people are bound up with one another, sharing despite differences, a common identity . . . that people participate in shaping the larger grouping of which all are members."[15] When Daly and Cobb speak of a "common identity" they are referring to both the economic and ecological processes that affect people's lives, and the decision-making processes that they commonly share. People form a common identity based on their shared commitment to the commons. This becomes the glue of community life, representing the merger of private and public interests.

For example, consider the different ways a person can perceive recycling. If it is merely a technical exercise, a consumer habit, a more politically correct way to throw out the garbage, then it only minimally contributes to community participation. It is merely a more efficient way to dispose of personal waste. However, recycling could also be the basis of a series of community interactions: educational programs in the schools, community rituals and habits linked to the recycling act, imaginative celebrations or other ways of joining people together, public meetings about the economic impact of the recycling process, or discussions about the ecological meaning of recycling. Through these actions and discussions, the community develops a shared awareness of the commons. In the case of recycling, the commons represents a community's understanding of what to do with their "collective" waste. By understanding that waste is an aspect of the commons, it is clear how the recycling process can have a long-range impact on many aspects of community life.

Increasingly, in the last few decades, diverse groups of people have formed community bonds as a result of their concerns regarding the ecological commons. They discover that what they have most in common is their membership in the ecosystem.

Daly and Cobb offer a working definition of community, explaining that there are many plausible scales of community size and configuration, and different groupings of people may be more or less communal depending on the degree to which they adhere to the following criteria: "A society should not be called a community unless (1) there is extensive participation by its members in the decisions by which its life is governed, (2) the society as a whole takes responsibility for the members, and (3) this responsibility includes respect for the diverse individuality of these members."[16] These are good criteria because they emphasize the importance of shared decision making and are applicable to a variety of configurations, including neighborhoods, electronic communities, regions, or even nations.

There are three sets of community patterns that weave through geographical, psychological, and political space, constituting what might be called *pluralistic community orbits*, representing the idea that a person may reside in many communities simultaneously. The combined impact of high mobility and electronic communication increases exposure to *circumstantial communities*, temporary groupings that form because people share the same space, attend the same event, or become members of a short-term network. Examples include conferences, backcountry campgrounds, or electronic town meetings. People can enter such arrangements as private individuals or with a commitment to community awareness, by thinking about the common needs of the group members. Some of these groups may congeal and endure, others may dissolve. They can function as transitional, cohesive communities, especially if all members recognize what they have "in common." Such arrangements will be as deep or as superficial as their members allow.

Second, there will continue to be, as described in *Habits of the Heart*, *communities of memory* in which people are linked through an oral or written tradition, which may include folklore, ethnicity, spirituality, geography, or other forms of kinship and ritual. There is a great risk, however, that twenty-first century demographics will make this prospect increasingly untenable. For the environmentalist, it is through ecological metaphor, the practice of subsistence, stories of working the land, the search for ecological identity, and the commitment to place that these communities of memory can be sustained. This is how the spirit of place and the notion of bioregionalism are most closely tied to community life.

Third, the idea of the commons necessitates the conception of *ecological communities*. This is the interconnected lattice of habitats and species, including human ecological practice, the impact that humans have on the earth, as well as the ways in which landforms, climates, soils, flora, and fauna affect humans. It is the recognition that humans have animal origins, including ecological and evolutionary links to the great biogeochemical cycles. Ecological identity broadens the concept of community so that it stretches beyond the limited sphere of human relations. A community of memory may articulate its tradition through its conceptualization of this ecological community. This includes metaregional communities that form around interconnected global and bioregional concerns, thus perceptually linking the local and the global (more about this in the sense-of-place map discussion in chapter 6). What people have in "common" with other members of an ecological community is the process of living together in a habitat. The path to ecological citizenship necessitates forging a community decision-making process regarding the land, air, and water that people use in common.

Both environmentalism and citizenship revolve around the commons. Environmentalism informs citizenship by placing an ecological perspective on notions of property, money, community, land, and power. Citizenship informs environmentalism by placing a mutual, participatory, and public orientation on human-nature relationships. Thus environmentalism and democracy are intrinsically linked. Environmental problems cannot be solved unless citizens formulate a suitable politics of the commons, as well as modes of participation that allow people to become engaged in public decision making. In order to do this, people must learn how to balance the tensions of democracy, such as property and the commons or the individual and the community, placing those tensions in a coherent ecological and political perspective. How can this be achieved within everyday life, making community participation both challenging and fruitful, a cleaner more appealing arena, something that all citizens have in common? How do we reconcile the tensions of democracy and begin to forge a meaningful politics, an approach to public life that is neither naively optimistic nor hopelessly cynical, that allows the negotiation of common interests in an inclusive and participatory realm?

The Commons Is a Place

From the literature of environmentalism and the experience of some innovative political figures, a politics of place emerges, a way for people to discover what they have in common by looking at the places in which they commonly dwell. This section reviews three interesting perspectives on place, participation, and the commons: Daniel Kemmis suggests that political culture can be revitalized through a bioregional approach to communities and neighborhoods; Vaclav Havel adds the moral ingredient; and Elinor Ostrom explores traditional arrangements to common property resources. Their work demonstrates some of the conceptual and practical potential of an ecological and moral orientation to citizenship.

Daniel Kemmis, in *Community and the Politics of Place,* examines how the "strengthening of political culture, the reclaiming of a vital and effective sense of what it is to be public, must take place and must be studied in the context of very specific places and of the people who struggle to live well in such places."[17] His book is particularly useful because it is such a balanced application of theory and practice. Kemmis presents some intriguing perspectives on political theory, gained through the rough-and-tumble real world of politics. He was the minority leader of the Montana House of Representatives and is presently the mayor of Missoula. It is appropriate to look at his ideas in some depth.

Kemmis is interested in what he calls the revitalization of public life, how to create spaces of political interaction that allow people to solve their problems together. He is concerned that people perceive themselves as split between two unworkable poles: the independence of rugged individualism and the leviathan of regulatory bureaucracy. In the American West, these approaches arise from the geography of the landscape and the history of the frontier. People relied on their frontier individualism in the face of a difficult and harsh environment, yet parts of the American West always remained in the public domain, subject to the decisions of a large and distant federal government. Rugged individualists see the bureaucracy as a threat to their ability to use the land. But for many, unfettered individualism poses a threat to land and only regulatory policies can preserve the common good.

These clashing approaches to public land often result in political stalemate, the appearance of intractable positions and irreconcilable special interests. People are placed in conflict situations and asked to choose between, or balance, a variety of competing policy alternatives, none of which seem palatable. Public hearings, for example, typically ask opposing parties to advocate their position before some adjudicating body. Kemmis suggests that politics is most effective when people are committed to a cooperative process that allows them to find a common ground. But the problem remains: how do you construct a political process that allows such cooperation to emerge? Kemmis is concerned that the "valley of common ground remains hidden" because in most public arenas values are considered private and subjective. Rather an "effective form of public life can only happen if we can learn to say words like *value* in the same breath with words like *public* or *objective*."[18] People must be willing to say what they think publicly and to forge a consensus from a diversity of perspectives.

For Kemmis, an authentic political process will emerge when people understand how they learn to inhabit a landscape together, how to explore the values about a place they have in common, and how they are jointly engaged in the cultural habits and material practices of everyday life. So he suggests that people "reinhabit" their community landscape and transcend the individualistic notions that drive them apart. Most of us assume that "the way we live in a place is a matter of individual choice" and we "have largely lost the sense that our capacity to live well in a place might depend on our ability to relate to neighbors."[19] This leads Kemmis to the work of Gary Snyder and Wendell Berry, who have written exemplary essays on how people live together in places, how the land becomes a tapestry of ecological and political work, and how people who coexist in a habitat form common bonds through the process of working together.[20]

Kemmis describes an interesting situation in Missoula. An environmental group and a logging company had different perspectives on whether mill wastes could be discharged into a local river. They were skeptical about whether a public hearing would resolve the issue. So, step by step, they carved an alternative, face-to-face, collaborative political process, attempting to forge a decision that would accommodate both sides. They presented a joint solution as common decision makers. This was only possible because they achieved a conceptual breakthrough, understanding their joint stake in the economic security and environmental quality of the community. They needed to make long-term plans with the commons in mind, that is, the river, the people, and everyone who had a stake in the outcome. For Kemmis, this is a model for how the politics of place can reinvigorate public life.

> But what holds people together long enough to discover their power as citizens is their common inhabiting of a single place. No matter how diverse and complex the patterns of livelihood may be that arise within the river system, no matter how many the perspectives from which people view the basin, no matter how diversely they value it, it is, finally, one and the same river for everyone. There are not many rivers, one for each of us, but only this one river, and if we all want to stay here, in some kind of relation to the river, then we have to learn, somehow, to live together.[21]

Before people can become citizens, they must see themselves as neighbors. It is their attachment to a place, the fact that they all live there and care about it that brings them into relationship with one another, making them neighbors. This is the basis on which "places may play a role in the revival of citizenship." People must learn how to live together and develop the trust, cooperation, toleration, and sense of justice that allows this to happen. Kemmis emphasizes the need for nurturing and cultivating the ways people work, play, and grieve, patterns of life that can be designed to enable people to live well in a place. These activities nurture the foundation for public life. He looks for examples of what he calls "practices of commitment," events which are not ostensibly political but in which diverse members of a community (neighbors) join together for a common purpose. Such events may include softball leagues, neighborhood watch programs, volunteer fire departments, or 4-H clubs. In each case, people join around something that is concrete, a process of cooperation that improves the quality of life for all.

Kemmis understands how difficult it is to accomplish this neighborly citizenship. People might prefer to have different neighbors, but if they wish to stay where they are they must find ways to live together. In many instances, it is easy for people to just walk away from neighbors who are unlike themselves, or who have different values. Perhaps it takes less effort to escape to the confines of privacy. But if people have no other choice, if they understand how they are tied together through the commons, then they will find a way to live together as neighbors in a vital community. Kemmis describes this common ground as a "high, hidden valley which we know is there but which seems always to remain beyond our reach."[22] The hidden valley is the place of shared values.

Similarly, citizens must reevaluate the economic structures of their communities: determining how much community wealth and livelihood is tied to larger, multinational interests that operate outside the region, how the community can maintain the integrity of its local economy, how to derive a politics of appropriate size and scale to match regional needs, keeping the forces of international economics and politics in perspective, managing the interface and cooperation between regions. These are some of the issues that are central to a political economy of place and that form much of the agenda of contemporary environmental politics.

Vaclav Havel grapples with similar questions in *Summer Meditations,* a series of essays in which he reflects on his political philosophy, specifically commenting on his vision for the former Czechoslovakia. Havel is mainly concerned with the morality of politics—how to forge a political process that brings the best out of people and that allows them to find common ground, especially given the uncertainty of their everyday lives. He wonders how a nation can become a community, given the diversity of regions, perspectives, and ideological dispositions. As a political leader, where can he find the common ground? He claims that a common politics is not possible unless we search deep within ourselves and discover the "moral origin of all genuine politics" and in so doing rediscover and cultivate a "higher responsibility" to serve those around us, serve the community, and serve those who will come after us.

Havel elaborates an ecological, bioregional vision which stresses decentralized decision making, with semiindependent principalities and well-defined regional identities. The neighborhood becomes the center of cultural, political, and economic life. Havel writes, "Small communities will naturally begin to form again, communities centered on the street, the apartment block, or the neighborhood. People will once more begin to experience the phenomenon of home. It will no longer be possible, as it has been, for people not to know what town they find themselves in because everything looks the same."[23]

Havel describes pluralistic networks of processing and marketing cooperatives, the ecological restoration of damaged landscapes, the transformation of agriculture, and a range of policies and ideas which will contribute to an ecologically oriented politics and economics. For Havel, this should take place in a market economy because "it is the system that best corresponds to human nature."[24] But this approach will fail, he warns, if it is not held together with a strong morality based on the higher responsibility of the commons. This is the fabric which holds together his conceptions of neighborhoods, civic culture, and political process. It has been difficult for Havel to convey this vision in practice, as powerful market forces and a resurgence of ethnic identity have become the driving themes in Czech and Slovak politics.

What evidence is there that people can manage the commons? What are the political contingencies of a politics of place? This question has received a great deal of academic attention, especially among political scientists, who are interested in understanding institutions for collective action. Elinor Ostrom, in *Governing the Commons,* provides an excellent study of the various frameworks used for analyzing and interpreting the management of what are called "common property resources." She describes the variables of scale and habitat in which these situations occur, ranging from small neighborhoods to the entire planet. Although most people conceive of policy solutions based on normative political structures, that is, central regulation, privatization, or regional regulation, there are other alternatives that provide interesting approaches.

> What one can observe in the world, however, is that neither the state nor the market is uniformly successful in enabling individuals to sustain long-term, productive use of natural resource systems. Further, communities of individuals have relied on institutions resembling neither the state nor the market to govern some resource systems with reasonable degrees of success over long periods of time.[25]

Ostrom analyzes what she calls "long-enduring, self-organized, and self-governed CPRs" (common property resources). She elaborates specific "design principles" which are illustrated by these long-enduring CPRs. Although it is beyond this book's scope to present all of these principles, there are several which are highly relevant. First, she notes the prevalence of "collective-choice arrangements," situations in which the individuals who are most affected by the rules governing the resource have sufficient inclusion in decision making about the resource. Second, she notes the prevalence of "conflict-resolution mechanisms," which are respected for their fairness, inclusiveness, and efficiency. Third, she cites the "minimal recognition of rights to organize"; local actors can make decisions without the interference of external governmental authorities. Fourth, she emphasizes the importance of understanding the local ecological and geographical conditions, insuring that various CPR rules conform to the local ecology and economy. Ostrom is hesitant to draw conclusions about the universality of these principles, but she is confident that much can be learned about CPRs by studying these examples.[26]

Kemmis, Havel, and Ostrom provide three interesting approaches to ecological citizenship. Kemmis demonstrates the importance of a politics of place, pointing out that such a politics is derived from the bioregional ties that bind a community. Havel emphasizes the moral responsibility that must pervade a sense of community and civic culture. A bioregional politics demands this moral fabric, the understanding that individual actions must serve a larger community. Ostrom analyzes the political rules that are necessary for successful management of the commons, demonstrating examples of cultures that manage the commons democratically.

But given the scale of global environmental change, the reach of international economic and political forces, and the predominance of the morality of the market and the bureaucracy of the state, it is difficult to imagine how a local politics of place will solve complex global problems. How can strong collateral and international agreements regulating the use of the biosphere emerge from a bioregional politics of place? And if a politics of place is the key to a new kind of higher responsibility, what kind of politics will emerge when electronic places become the setting for so much political dialogue?

If electronic communities have the potential to make people global neighbors, perhaps there are principles of ecological citizenship that can be applied in different settings. As the river is the focal point for the residents of Missoula, Montana, the biosphere could become the focal point of a global neighborhood. The perception of the commons is also a function of scale. One's ecological identity isn't limited to the local circle of a bioregion, it includes the global biosphere. The planet consists of a complex system of hundreds of integrated ecosystems and bioregions. Hence the concept of neighborhood is irrevocably expanded. Perhaps it is no harder to make political decisions in one's proximate community than it is to arrive at a protocol for banning the use of chlorofluorocarbons (CFCs). In both cases, decisions cannot be made unless people perceive themselves as neighbors. The scale of the neighborhood is vastly different, but the principles governing the commons are remarkably similar. In both cases, it is the perception of the commons that is important. It is the awareness that one's neighbors are both next door and halfway around the world.

In this way, the formation of ecological identity contributes to local and global ecological citizenship. By expanding the circles of identification, the global commons becomes the local neighborhood. Whether this process occurs at the level of the psyche (breaking down the barriers between the ego boundary and the ecosystem) or the polis (creating neighborhoods in which people can develop a civic culture) or the ecosystem (understanding the impact of one's actions on the commons), a citizen is obliged to balance individualism and the community, private property and the ecological commons.

This is a perennial struggle, a lifelong task. It may take generations before such attitudes prevail. But this doesn't mean such activities are fruitless or impossible. Havel discusses the overwhelming circumstances of this work:

> A heaven on earth in which people all love each other and everyone is hard-working, well-mannered, and virtuous, in which the land flourishes and everything is sweetness and light, working harmoniously to the satisfaction of God: this will never be. On the contrary, the world has had the worst experiences with utopian thinkers who promised all that. Evil will remain with us, no one will ever eliminate human suffering, the political arena will always attract irresponsible and ambitious adventurers and charlatans. And man will not stop destroying the world. In this regard, I have no illusions.[27]

Havel warns us that this is not a battle between good and evil, or the honorable and dishonorable. Rather, it is a process that takes place inside ourselves. We each must strive to find a higher moral ground. This is what it means to be human. Just making this effort does not ensure that the world will then become the place we want it to be.

But we can develop the learning tools and the practical skills that allow us to understand the political struggles of our place, and how environmental issues require direct political attention. Through ecological identity work one can bring these issues to bear on personal decisions, understanding how ecological citizenship is deeply embedded in the practice of everyday life, integrating the experience of nature with a vision of political action. Widening the perception of the commons is one means of doing so.

Nevertheless, environmental politics is the terrain of harshly fought battles in which powerful interests have a great deal at stake. On the individual level, ecological identity work can inspire insight regarding community participation and the perception of the commons, laying the groundwork for ecological citizenship. But the process of politics can be daunting and intimidating. One must also prepare for the hard work of politics, the reality of power and controversy. In the public domain, where people have vastly different values and interests, it takes a great deal of inner strength and commitment to forge consensus and community, and in some cases difficult conflicts emerge. Chapter 4 examines the process of citizenship and the quality of participation, linking ecological identity to political identity, exploring the learning spaces that allow people to reflect on their political experiences.

4 Political Identity and Ecological Citizenship

For several months I have collected all of my environmental and political junk-mail in a large box. It is the first day of a course on ecological citizenship. I lug the box to class and dump its contents in the middle of the floor, asking the students to sort through it. "What does all of this material reveal about environmental politics?" I ask.

After the predictable environmentalist lamentations about all of the felled trees that contributed to this pile, the group settles in to discuss many of their frustrations about political activism. One woman describes her busy life as the director of a small environmental science consulting firm and the mother of two young children. The only way she can enter environmental politics is to send money to "effective" organizations. But another woman challenges her, claiming that such actions are ultimately superficial. She complains that she is besieged by fundraising requests, through the mail and the telephone, from people who read rote messages and are uninterested in meaningful dialogue. She describes this as the advocacy marketplace, comparing mail-order environmental activism to a home shopping channel. "I want dialogue, exchange, and genuine sharing of diverse opinions," she says, "not a quarterly payment to another demanding group."

This becomes a stirring debate, getting to the source of what I detect is a profound ambivalence regarding political participation. One man explains that political involvement is taxing and discouraging: the endless meetings, strategy sessions, and public forums. To enter politics, he claims he must learn how to manipulate, cajole, or wheel and deal, expressing discomfort in these roles, but knowing that he must "play those games" to accomplish anything. He describes environmental politics as "the bottomless pit of the public arena." Once he gets in, it's hard to get out, and he's not sure what he accomplishes anyway. "Environmental politics is the province of the professional activist," he says.

An environmental educator surveys the colorful brochures on the floor and declares that he is overwhelmed by all of the causes. "Look at all of the problems described here—boat people, toxic waste, endangered species, global warming—what difference will my small actions make anyway? Why should I bother to get involved in politics when I can have a more rewarding experience turning children on to nature?"

Another student chides him respectfully, describing his condition as the Pandora's box syndrome, suggesting that he is afraid to open himself to the suffering these issues entail, fearing that once he becomes involved, the emotional and moral ramifications will cascade into his comfortable life. "Why must political activism be perceived as so black and white? What's wrong with a strategic, limited involvement that reflects your interests and temperament?" The man responds that in his own way, working with kids in an elementary school classroom, he is contributing to the common good. "I live an ecologically sound lifestyle and I don't create all of this meaningless junk mail."

Finally, several people suggest that our conversation perfectly recreates their internal dialogues. They feel compelled to act as responsible ecologically minded citizens, but they are not exactly sure what this means. They want to become involved in their communities, and help make decisions about the commons, but the conditions of participation seem inaccessible and daunting, or they are overcome by the demands of their professional and personal lives.

There is another side to this discussion. People also describe the positive quality of their political experiences, those that are tied to cooperative ventures, coalition building between diverse groups, or bridging conceptual gaps that lead to broader understanding. Others talk about the many faces of environmental activism, that it need not always be tied to traditional politics, that one can take smaller actions on a local scale. Most of them have some kind of political success story in which they act on their values, helping to implement some type of change. Yet still, a sense of doubt prevails. Will political actions make any difference, given the scale of environmental problems?

As an educator I am interested in developing approaches to ecological citizenship that allow political participation to become a personal and collective learning experience, ways to understand power and controversy as tools of awareness and transformation. The challenge is to design learning spaces that encourage people to understand how they perceive power, how they can develop an effective and appropriate political voice, and how political participation can become an accessible, fulfilling, and fruitful experience—a means for the public expression of ecological identity.

Ecological citizenship hinges on a crucial conceptual step, the integration of ecological identity and political identity. In chapter 1 I described the reflective processes that facilitate ecological identity, the learning experiences that constitute an ecological worldview—a sense of belonging to a larger community of species, an understanding of the ecological commons, the broad ecological impact of personal actions, how people identify with nature and ecosystems. These experiences also take place in social systems. Inevitably, when people work together, share the same habitat, and make decisions about the commons, relational issues of power and controversy emerge. Ecological identity emerges in a social and political context.

Politics describes the public process through which people resolve contested perspectives or interests. *Political identity* refers to how an individual perceives his or her power in relation to other people, groups, and institutions. *Political identity work* is a series of reflective approaches that enable people to reconceptualize the role of power and controversy in their lives: how issues of authority, conflict, and consensus are intrinsic to their sense of self and define their participation in a social system. Just as ecological identity work is intended to broaden and expand the circles of ecosystem identification, political identity work is intended to broaden one's perception of the circles of *relational power*, enabling people to see how power permeates all aspects of everyday life—not as an instrument of control, but as a means to expand choices; not as something external to a situation, but as intrinsic to an inclusive decision-making process. Power, like the commons, is what one perceives it to be. Political identity work emphasizes both the complex chain of relationships that emerge from personal actions and the processes through which collective decisions get made.

Power is considered "relationally," representing the capacity to expand mutual choices and collective decisions, as opposed to a zero-sum approach, in which one person's power gain is seen as another's power loss. Hannah Arendt's definition is instructive:

> Power corresponds to the human activity not just to act but to act in concert. Power is never the property of an individual; it belongs to a group and remains in existence only as long as the group keeps together. When we say of

somebody that he is "in power" we actually refer to his being empowered by a certain number of people to act in their name.[1]

Some theorists distinguish between *power over* and *power with.* "Power over" is described as coercive, an instrument of domination, in which one side gains an advantage over the other. "Power with" implies that people develop their choices and objectives by acting together, or what Seth Kreisberg describes as relationships of coagency and the language of assertive mutuality.[2]

The purpose of this chapter is to explore how political identity unfolds in an educational context—leading to a vision of ecological citizenship, formulating a political voice for one's ecological identity. The first section explains the concept of political identity work in more detail, showing how the use of memory and the reinterpretation of political life experience can lend great insight into one's political values and motivations. Political autobiography combines family dynamics, historical circumstances, and a review of one's "political" past as crucial aspects of political identity work. The second section considers "power flow analysis," a series of learning activities that trace the complex flow of power in controversial situations, exploring how power and controversy can be used as learning tools. The third section develops a practical concept of ecological citizenship through the integration of ecological and political identity.

Political Autobiography: A Cartography of Political Identity

In my teaching, I use a process called *political autobiography,* designed to help people understand their political values, articulate a coherent political identity, and translate this knowledge into meaningful political action. Political autobiography consists of a two-phase process. With the *political genogram* students construct a portrait of their family political history, as a means to record the political values of their grandparents, parents, and siblings. With the *political life-cycle chart* they recall seminal political moments in their life histories, tracing the patterns of political engagement in relationship to the developmental context of childhood, adolescence, and adulthood.

Both activities introduce political identity work. Just as ecological identity is formulated through cycles of human development and various experiences of nature, so political identity is formulated through one's changing perceptions of power, ideology, and history, as reflected in family dynamics and phases of personal growth.

For educational purposes, political identity can be organized around three interconnected dimensions: ideology, temperament, and action. *Political ideology* refers to one's political belief system: values regarding capitalism and socialism, free markets and collective structures, regulation and coercion, right and left, conservation and preservation, etc. *Political temperament* describes one's disposition, values, and behaviors regarding conflict situations: conflict avoidance or assertive provocation, negotiation or belligerence, accommodation or intimidation, cooperation or independence. *Political action* regards the activities of implementation: voting, civil disobedience, letter writing, organizing, or apathy.

These distinctions may be incident-specific. It's one thing to take a stand on a major international issue, it's another to participate in local politics, and yet another to tell your husband that he has to take more responsibility for child rearing. These are questions of scale and temperament. Where do you perceive conflict and when are you willing to take a stand? Issues of either personal or public conflict develop in extremely complex ways, and one's political actions depend on the context: how much is at stake emotionally, the extent to which values are compromised, the amount of personal effort required, how much risk is involved, and so forth.

Political autobiography considers the three dimensions of political identity for an entire family system. Typically, one's ideology is a mixture of cultural, historical, and family variables. Political temperament and action are significantly influenced by how families deal with conflict. Parents are the principal role models for conflict resolution. For example, many families shun discussions of controversial issues, yet they have complex power relationships. Political identity work relies on the use of memory: How were conflicts dealt with in the home? How did people respond to one another in controversial situations? What seminal historical events prompted deep reflection or difficult ethical decisions?[3]

Phase 1: The Political Genogram

The political genogram is essentially a variation on the family tree, in which grandparents, parents, siblings, and other important family members are symbolized with boxes containing specific information. A series of lines connects the boxes according to typical family tree relationships (marriages, etc.). Using several keywords, each family member is described according to the three dimensions of political identity (ideology, temperament, action) as in the following examples: ideology (conservative, liberal, radical), temperament (authoritarian, accommodating), action (union member, voter, uninvolved). A color code creates visual uniformity within the chart, i.e., ideology (purple), temperament (red), action (green). The descriptive words are placed directly in the box corresponding to the family member, or in a suitable adjacent position. Any important historical events are denoted in bold lettering. Finally, the author of the genogram describes himself or herself, using the same categories. I invite participants to use their artistic imagination in developing the diagram. Most important, it should be clear and attractive so the chart lends itself to interpretation (see figure 4.1).

Often people engage in preliminary research in order to be able to complete the diagram, calling family members to find out about grandparents, talking to parents and siblings about their impressions of one another, and so on. For some, this is a very difficult emotional activity, as it can bring up anxiety regarding their family. I make it clear that a person should only research and reveal what is appropriate. The assignment is designed to cultivate political insight, not family therapy.

I rely on collective interpretation, breaking a class into groups of four to discuss the genograms. What emerges is a splendid review of several generations of American history, with the full gamut of representation: labor organizers and corporate executives, immigrants and Daughters of the American Revolution, coal miners and professors, radicals and conservatives of all stripes. Students write comprehensive family profiles, noting how various family members influenced their political identity. They review the dynamics of family conflicts as a means to an understanding of their own approach to conflict.

Jane is a middle-aged educator, raised in South Africa, where she was surrounded by political turmoil. She realizes that her parents were mixed role models, setting different kinds of examples. She uses this activity to evaluate the "good and the bad" and to sort out the influences that are most important. For example, her father, although he was a strong and courageous critic of apartheid, had a low opinion of women and ran an authoritarian household.

My father had the biggest impact on my life due to his overpowering, irascible, and impatient nature, making him a harsh disciplinarian, who believed

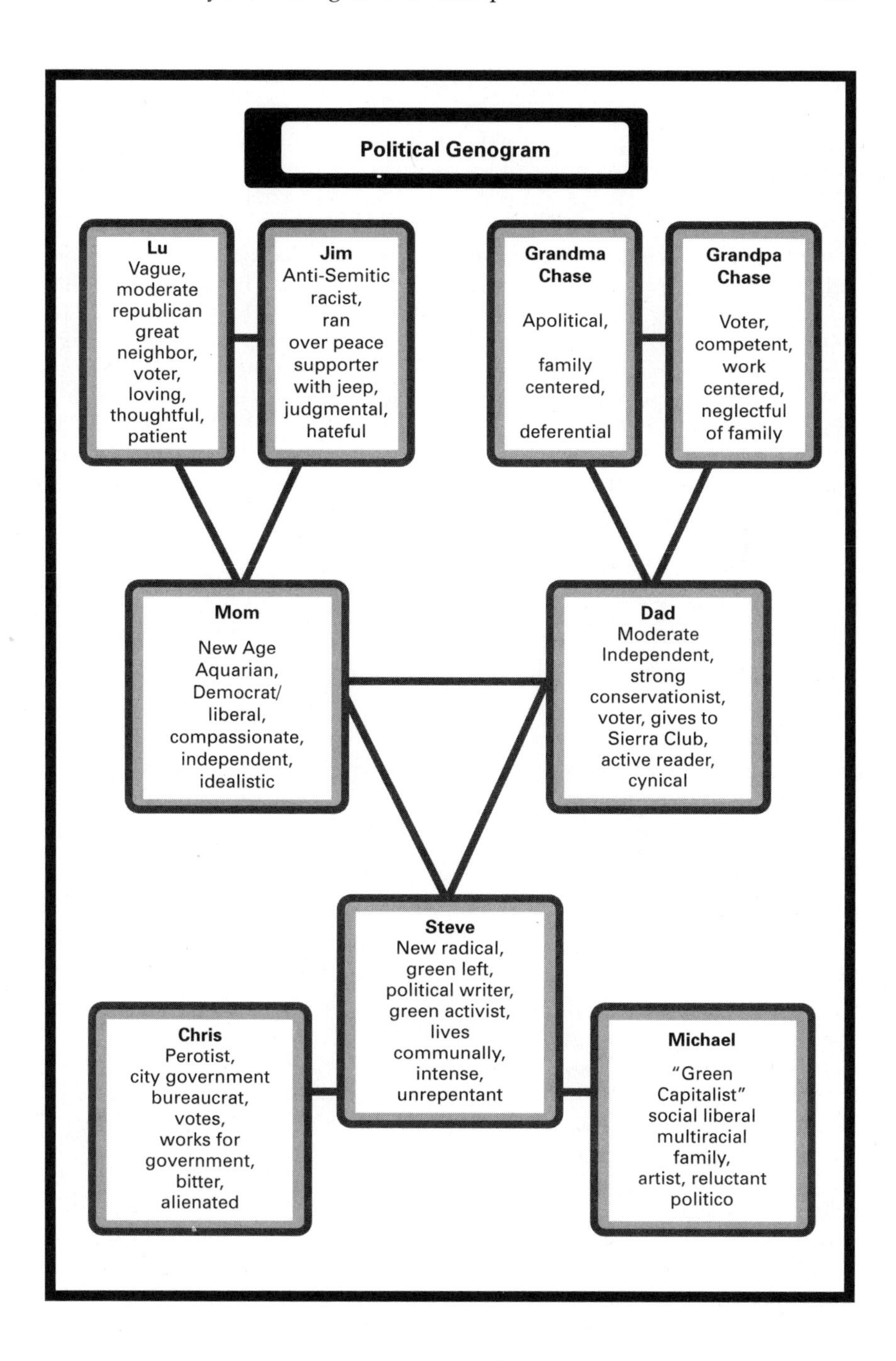

Figure 4.1
A political genogram.

that children were to be seen and not heard. He was a record-breaking bodybuilder, a practicing mystic following a strict path, and a dedicated businessman causing him to appear a formidable figure to a child. The fact that he tore down apartheid signs at work and was always arguing with nationalist colleagues meant he was rarely in a good mood, as he had to face being constantly undermined and attacked in his work. He lived his beliefs and tried to inspire hope in his many African workers. Sleeves rolled up, overalls over his suit, no task was too menial for him in order to motivate his employees. This was considered bizarre behavior in a culture where Africans served tea to whites in their offices and did all the menial work. He had a genuine concern for the welfare of his workers and their families, which meant he was hardly ever at home. He spent most of his time in the company of men and had a low opinion of women, which led me to want to be like a man and not want to associate with women's things. Because my home life (and country of origin) was so conflict-oriented, I tend to expect conflict rather than cooperation and assemble my army (support) in advance.

Jane received a contradictory message: it is important to have an appreciation of social justice and to act on your convictions, but it is not acceptable to explore your independent ideas. Jane realizes how so many of her ideas about politics have their origin in her childhood, and now, almost 40 years later, she is able to revisit those events, and place them in a perspective that conforms to her contemporary ecological and political values, giving herself permission to explore women's spirituality, to take on specific advocacy roles, and to approach conflict differently.

Charles, an environmental biologist, recognizes that although he and his father have very different political philosophies, once he looks through the superficial ideological issues, he can use his father as a role model for community service. He questions his father's conservatism and judgmental demeanor, but he appreciates the ways he contributes to community life.

My dad and I have rarely occupied common ground, even when we were sharing the same house. Simply stated, he is firmly entrenched in Republican Party ideology and often describes other viewpoints as devoid of worth. He argues the conservative opinion with great fervor and becomes loud and angry at the suggestion of alterations to, or variations or disapproval of, his convictions. Easily dismissing notions other than his own as claptrap, he rejects opportunities to engage in meaningful dialogue and condemns the values of liberal progressivism. He will then back his claims by spouting out the Republican party line du jour. Although he had no political ambitions, he became the mayor of his community (an all-white, Republican, and rich suburb) for a short while. On the other hand, my father taught me a great deal about the larger community. As a prominent business leader in a moderate-sized city in Tennessee he was a spokesperson for raising the standards of living in that community and was particularly inclined to sponsor athletic scholarships, and he was a recognized supporter of the local university. His company donated time and people power to civic projects throughout the region, and I think his instinctive efforts to enhance the lives of people in the small southern city speak much louder than his rather one-sided political affinity to conservatism.

Charles understands the roots of his passive-aggressive approach to conflict. He wishes to overcome this by becoming a better listener and dealing with conflict in a more balanced way.

Middle ground did not exist in our house. We lived in a world of intense conflict or passive avoidance. Perhaps this explains my habit of mollifying awkward scenarios. I've inherited a little of my mother's knack for accommodation and my father's tendency to be overbearing and intimidating, and I am honestly not satisfied or comfortable with either. As a result, I attempt each day to listen to others and understand the point of view of those with whom I do not agree. I firmly believe that this practice will lead to improved mediation and conflict resolution on my part.

Chris, a wildlife manager, recalls some of the difficult family dynamics that influenced her political identity. Her political genogram brings up many of her painful childhood experiences, the incessant conflict in her house, the difficulty in being heard. She realizes how her orientation to conflict comes out of her childhood and adolescence:

I have been most politically affected by my father and by the relationship between my mother and father. The dynamics of my parents' relationship never left any space for political discussions or intellectual talk at the dinner table. This stemmed mostly from my father's perception of my mother as being an idiot. It turns out that he was wrong. Now, years after her death, I am finally coming to know something of my mother. It is true that she was never terribly concerned with intellectual matters, but she did have a diversity of opinions which she was never shy about sharing. My father just happened not to agree with one word my mother ever said. Because my father is a very bright and articulate man, we three girls eventually learned to discount most of what my mother said and learned to sit silently (awestruck) as my father rambled on about what a fool John F. Kennedy was, why Nixon was a great guy, and the like. Of far greater importance than my father's dominating political views, however, was the dynamic created in our home life between my parents. My mother was unbelievably outspoken about everything and so every discussion in our house soon devolved into a LOUD discussion until it dissolved completely into a shouting match. Then someone would get disgusted enough, would walk off and slam a door. Conflict was part and parcel of my everyday life. While I never heard my parents champion a cause, except, of course, my father championing the Republican party line, I did grow up learning how to argue, how to disagree about nearly everything with my father, and was never afraid to voice my opinion.

As an adult, Chris takes an active role in her community. She realizes that her political style has been overly aggressive, that her relationship with authority is unresolved. With this awareness, she has been able to take a more compassionate role, even to the extent of moderating her town's public meetings. In her case, the parallels between understanding family dynamics and her behavior in the political sphere have helped her overcome some of the barriers to effective public participation. She tries to understand the conflict inherent in a situation and to defuse it, or transform it accordingly.

Stiles, a resource manager, grew up in a family that was apolitical. He appreciates the roots of his own inaction, wondering how he can extricate himself from the dynamics of political denial and apathy.

About the only thing I can say for myself and my family is that we are all basically conservative individuals who are hesitant to change. I believe that a large part of my family's ability to remain apolitical about most things is the fact that they have been relatively well off for most of their lives, and therefore somewhat insulated from the radical changes that the world has undergone during their lives. Without the luxury of having enough money to protect ourselves from the world's sudden changes, I believe that my family would have been forced to take a decidedly more active stance on some political issues. As it is though, my family has adopted a somewhat fatalistic attitude regarding most political agendas. My own problem is that I have now come to accept the fact that there are some things in the world that one simply cannot change, and I have given up even trying to effect any kind of change that would be beneficial to me. As a result, I have learned to avoid conflict when possible, or at the least, to find an acceptable compromise that defuses the situation before it can deteriorate into a battle between polarized viewpoints.

Stiles realizes that his political voice is that of a mediator, someone who can engender trust, defuse conflicts, and find common ground. He will never be an "activist"—it belies his temperament—but there are other ways he can become involved in politics.

Steve, a writer, editor, and environmental activist, understands that his activism is a blend of many family member characteristics, but also recognizes that there was a single driving influence in his political identity: his mother, who provided him with the support and awareness that strengthened his activist orientation. As a supplement to his political genogram, he submits a thank-you letter he wrote to his mother:

You gave me the values of a freedom fighter which have shaped my life. You explained racism to me and told me about the Holocaust. You explained your views about the Vietnam War to me when I came home from school spouting crap about the need to kill Communists. You told me about Joe McCarthy. You gave me feminist books when I was 13. You stood up for me when Grandpa said I shouldn't be reading Gandhi at my age. You stood up for me when I got kicked out of my junior high school civics class for calling the teacher a racist. You supported me when I cut classes to go to the college for ecology day. You shared ideas with me and listened carefully. For all of this I thank you.

Steve's mother serves as a model for his own parenting—the need to provide political guidance and support, to allow his children to develop their own voices, and to listen to what they have to say. Perhaps this is more valuable than any books he can give them or ideas he can explain, the virtues of respecting his children's independent voices and strong opinions.

From my observations of several hundred political genograms, the following themes emerge: (1) How families deal with conflict and controversy significantly influences how children approach political situations, how they view authority, their willingness to engage in consensus, and their attitudes toward political problem solving. (2) Although parents' ideological predispositions have some influence on their children, other factors, such as attitudes toward community service, respect for other peoples' opinions, standing up for one's values, and open-mindedness, are the critical role-model influences on political identity. (3) Many families exhibit inconsistency between their professed political values and the way they make family decisions. The political process through which family decisions get made is more vividly recalled than voting habits or political opinions. (4) Adult learners use the genograms to extract the best qualities of their family members and synthesize them to support their emerging political identity. They don't dwell on the pain and anxiety, or use those as excuses for political inaction. Rather, those dynamics spur people on to understand how they can create more constructive political situations. The purpose of the political genogram is to better understand the roots of political identity, and to apply what is learned.

Phase 2: The Political Life-Cycle Chart

The political life-cycle chart is a simple graph which plots five themes across a horizontal dimension: age, major historical or personal events (black), political ideology (purple), political action (green), and political temperament (red). In each case, the graph represents a chronology, depicting important events, incidents, and so forth. For example, in elementary school a person doesn't necessarily have a political ideology, but he or she may have certain key values. As an adult, those values may have been transformed into a coherent belief system. During different stages of life, people may discover different approaches to political action. People become more conservative, moderate, or radical as they grow older. They grow more assertive or more accommodating. Or perhaps, a particular historical event serves as a watershed. The chart is projected 10 years into the future, challenging students to consider how their political identity is evolving, and what they aspire to achieve.

Children, adolescents, and adults approach political questions very differently depending on a range of circumstances: family background, cultural and historical situations, maturity, education, and so on. Hence one's perception of power and controversy and one's ability to act on public issues is highly variable depending on all of these factors. The purpose of the political life-cycle chart is to trace the formation of political identity according to one's personal development.

I advise people to search their life histories by attending to the following types of questions:

- In what circumstances do you act in accordance with your values?
- What motivates you to take action on public issues?
- When do you feel most politically empowered?
- When do you experience political apathy?
- What political actions cause you to be frustrated or to feel fulfilled?
- What do you consider to be worthwhile political participation?
- Are there institutions that facilitate this?
- Are there modalities of civic participation that best reflect your political temperament?
- What does this tell you about how you should become involved in the future?
- What would you consider as your first political act?

Katrina, a naturalist, recognizes that her first political act was prompted by spontaneous anger and outrage, yet it was closely connected to her ecological identity. She recalls what happened when a developer placed *No Trespassing* signs on a tract of land that was especially important to her.

I removed *No Trespassing* signs from the "front acres" of our house. My parents chose to send us kids to college rather than buy the 4 or so acres in front of our house. When the developer started clearing the land, he put up signs. I nearly flipped. I was furious. I had walked to church by myself and came home feeling peaceful and happy, only to be confronted with orange *No Trespassing* signs on land I considered my front yard. Without a second thought I ripped the signs down, aching inside that the man had NAILED the signs to the trees. A more horrible affront I could not imagine. Nails in a living tree! I never told anyone that I had taken down the signs. Luckily, the signs did not return. Perhaps my point was taken.

As Katrina interprets her political life-cycle chart, she notices an important pattern, how when she acts from her convictions she gains confidence and strength.

When I feel very strongly about an issue—abortion rights, injustice, inefficiency, not selling meat at the coop, drug testing, the Gulf War, the US support of civil war in El Salvador, etc.—even if my position is unpopular (I was one of a very few coop members and one of two board members who opposed selling meat), uncomfortable, or radical (or at least different), I go for it whole hog. I do not back down once my grip is set. Scathing letters, personal boycotts, protest votes, conversations or arguments with friends—I live my convictions. The greatest revenge is living well, or correctly, true to my beliefs and passions, true to myself. I am intensely unhappy if my life or a portion of it is not in sync with my passions.

Amy, an environmental science teacher, describes how she rarely got involved in political issues until her last year at college. In reviewing the formation of her political identity, she realizes that all of the experiences prior to that point were important steps, phases of learning and listening.

As I look back I recognize that seeds were being planted for my political identity. The radical young new minister at our church who was always talking about peace and justice (topics that were shunned by the congregation), the bilingual students I tutored in English in fifth grade, the nuclear power plant being built across the river from my city, the afternoons helping my mother sort canned goods in the food pantry.

But in her last year of college, Amy discovers the most crucial aspect of her political identity:

I came out as a lesbian, and suddenly my entire life was political. Much of my political activity since then has focused on gay issues, whether through coming out in the local high school where I was working, serving as an advisor to the lesbian/gay support group at the college where I currently work, or working for the addition of sexual preference to the New Hampshire antidiscrimination bill. I'd say the most actively political thing that I'm doing these days is singing. That may seem strange, but I am a member of a lesbian chorus. We sing music about civil rights for all people: peace, justice, strength, feminism, racism, and, of course, homophobia. Even without the political content of our music, just standing up in front of an audience and identifying ourselves as a lesbian chorus is an act of courage and a political message.

Barbara is an organic farmer and environmental educator, who has worked in Latin American and Europe. She never saw herself as "politically active," yet clearly, as she provides the details of her political development, her life experience is rich in politically, ecologically, and socially motivated actions and choices. She uses the political life-cycle chart to project her work into the future.

I am drawn to do seed work, planting seeds which will grow into the future. It requires a long-term view of the whole and a patience, sense of humility, and faith in much that is unseen. Seed work may or may not take as its task, usually not, the deposing of the political demons of the time. In my case it does not. When looking at my political life-cycle chart, most of it appears to be preparation for something yet to come. My intuition is that life tasks are preparing me for events that will happen when I am in my fifties to seventies. Creating centers of learning that offer the inspiration of experiencing in action new ideas and forms is what makes me feel most empowered. It is both the source of my personal transformation (bringing to life what lies inside me) and the source of inspiration for others. This is my political path.

She claims that her self-image changed as a result of this exercise.

I continue to be surprised at how I am perceived by others, especially others for whom I have respect. I went into class with a feeling that I have fallen short of my responsibilities as a citizen. I was surprised to hear from the others in my group that they thought I had accomplished a lot. It is a great moment in one's life when one's actions of dissent and nonconformity, which had hitherto been simply what one was compelled to do, are supported and confirmed by those for whom it is given to read the signs of the times: past, present, and future.

Eric, a high school biology teacher, assesses the various tensions that make up his political identity.

As I look at my political life history, I see four major areas of tension that are important in framing the way I act politically: first, there is my idealistic utopianism versus a realistic acceptance of the power and inertia of the status quo; second, fear of conflict derived from my psychological history versus my strength of self-expression; third, individualism versus communitarianism; fourth, my desire to minimize my personal suffering versus my willingness to accept personal suffering in my work in the world. All of these tensions are still very active in my life, and I'm sure they will remain so in the future. By becoming more aware of them, hopefully I can further work to integrate my political behavior within my holistic sense of self.

The most revealing aspect of these charts is that although many of my students do not see themselves as "political" activists, they have an extraordinary depth of political experience. This becomes particularly clear as they reconceptualize politics to include a range of power situations, not just voting or traditional advocacy roles. For example, Barbara, the organic farmer, has chosen sustainable agriculture as a career for political as well as ecological reasons. Her approach to environmental education is based not only on her ecological identity (the strong sense of identification with the land) but also on her political identity (her sense of justice, fairness, and community). It is impossible to separate these aspects of her self.

Other people recall how ecological and political motivations spurred public action: organizing recycling bins in high school, joining local conservation commissions, convincing their organizations to use recycled paper, running public programs on controversial environmental issues, teaching controversial issues in the classroom, taking unpopular stands at family gatherings, and so forth. In each case, they were motivated both by their ecological worldview and their sense of civic responsibility.

The political life-cycle charts can also serve an empowerment function. For example, one man described how he quit his school play in the sixth grade, because his teacher made unfair demands on the class, treating students differentially. In retrospect, he realizes how much personal strength he needed to take such an action. And his parents, who were very active politically, couldn't understand his actions, although if he had gotten up in front of the class and made a speech about disarmament they would have been delighted. They didn't understand that quitting the play was his political statement. So he acted without support. As an adult, he can review the political dynamics of the situation and recognize that his political voice emerges in many different ways.

The historical element is of great interest. Many people realize the extent to which they have been affected by historical events (the

Kennedy assassination, Woodstock, Chernobyl). When people consider how these events produced strong reactions and begin to link the patterns of their emotional and political responses, they get a better sense of how they participate in history. The political life-cycle chart, as a developmental portrait, shows how political identity is the flip side of national and international events.

Political autobiography is a lifelong educational process. Political voices evolve as people gain more political experience and better understand their relationship to power, the meaning of historical events, and how they want to participate in public life. In a sense, political autobiography links people to their family commons. When the genograms and life-cycle charts become the basis of public discussion, people see how their lives are integrated with a larger community, how political actions are critical to their sense of self, and how their ecological identity is inextricably linked to their political voice.

Power Flow and Political Identity

The second phase of political identity work is the ability to interpret the flow of power in political situations, and to strategize appropriate interventions, that is, ways a person or group can take actions to achieve a specific result. This may include advocating an environmental policy objective, making efforts to bring about a more equitable distribution of power, considering the ethical implications of power, or using power to transform controversy into mutual understanding.

This section offers four approaches to political identity work, each based on a form of power flow analysis. These include the power flow chart, a learning activity that enables people to interpret the patterns of power in controversial situations; a discussion of the everyday life experiences that can serve as a laboratory for political identity; a look at the use of language as an effective way of understanding power flows; and a review of the ethical dilemmas involved in the use of power in advocacy situations.

The overriding themes are the educational approaches that elucidate one's perception of power and how those perceptions inform political identity. I work from three basic assumptions: (1) people improve their capacity to resolve conflicts when they look deeply at the attitudes and motivations they bring to a political situation, (2) upholding a "power with" morality is as important as the urgency of environmental reform, and (3) controversy is the educational means through which people find higher purpose and common ground.

The Power Flow Chart

I ask my students to thoroughly analyze a controversial issue by creating a detailed chart, exploring the various ways that power flows through the issue. I interpret controversial issues broadly, to include national or international environmental politics, local community issues, organizational dynamics, workplace incidents, or even interpersonal situations that have public ramifications. What is most important is that a person really cares about the issue he or she chooses. "Power flow" refers to all of the actions, texts, statements, perceptions, and symbols that influence the situation and will affect the political outcome. The point of the activity is to break power down into its constituent parts, to see where it comes from, how it is attained, in what ways it emerges, and how it is defined—in effect, to demystify power relationships by organizing them in a systematic and thoughtful way. By using a chart or illustration, the power flow is rendered vivid and complex.

Bruce, an environmental educator and science teacher, carefully investigates a controversy regarding the expansion of a small airport. At stake is a row of trees which are scheduled to be removed. It appears that there is no compromise here—the trees either go or stay. His diagram, "Keene Airport vs. Swanzey Property Owners," traces the detailed sequence of events that led to a political stalemate. Essentially, the Keene Airport wants to expand and to do so a row of trees must be cut down. Local property owners are enraged by this, fearing not only the environmental consequences but also the potential depreciation of their property.

In figure 4.2, Bruce denotes the "players" as single line–bordered boxes and the "contextual influences" as double line–bordered boxes. The arrows represent how power is exerted. The "power actions" are numbered sequentially according to the chronology of events. The contextual influences (such as economic considerations) exert enormous power over both sides. He describes the media (*The Keene Sentinel)* as a player, in that it serves as a tool that can be utilized by the adversaries, and contributes a feedback loop of information exchange. He assigns all but one of the players to two groups: either "the trees

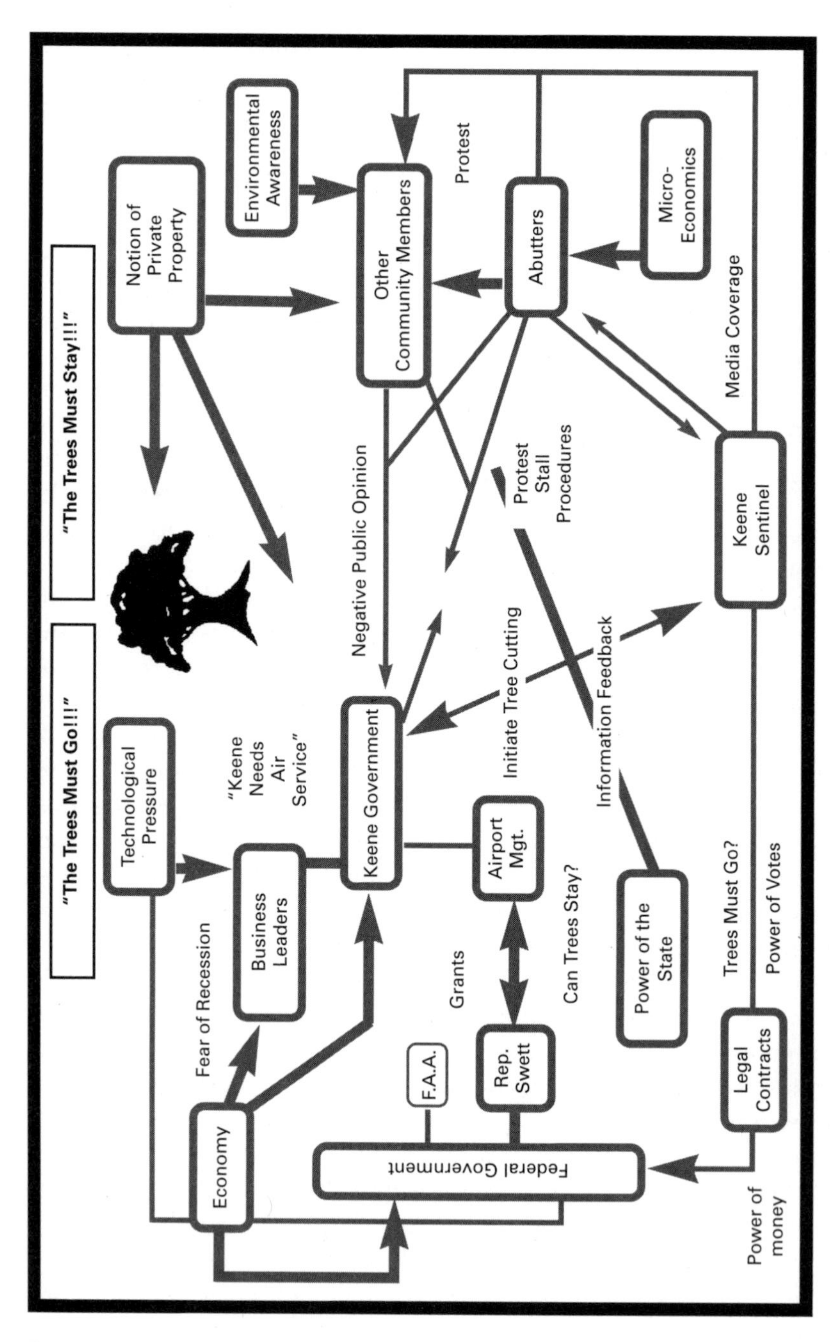

Figure 4.2
Power flow analysis: Keene Airport vs. Swanzey Property Owners

must go" or "the trees must stay." There appears to be no room for compromise. In the end only one side can gain the desired outcome.

Bruce concludes that the power on "the trees must go" side far outweighs the power on "the trees must stay" side. The latter group is comprised of a few residents in a small, rural community who are up against a hierarchy of power that leads to the federal government. It is only a matter of time before this much larger power will ultimately succeed. This is a "power over" situation. Nevertheless, the Swanzey tree group keeps the fight going much longer than anyone expects, using various legal means to delay the process.

Bruce believes that indirect, implicit, ideological considerations carry the most influence in the long run. These contextual influences are evenly distributed between the opposing groups and reflect the tensions of contradicting values in a liberal democracy. These tensions set up a kind of equilibrium within the conflict, preventing a quick outcome. He perceives three tensions: (1) the power of the state vs. the power of the individual, including private property vs. eminent domain, (2) regional economic growth vs. individual economic welfare, and (3) the notion of justice based on principle vs. legal contract. Bruce has almost a fatalistic view of this situation. The opposing players are trapped in their roles, imprisoned by their values, and linked to a larger ideological context.

What does Bruce conclude from this? He assumes that there are many political situations that are so complex that a short-term compromise appears impossible. In this case, the trees represent the ecological commons, yet they are merely pawns in a regional economic dispute. He suggests that this particular issue also exists in a larger context. For example, it is one of many regional environmental-economic issues. From a strategic perspective, the environmentalist, or any political group for that matter, may have to trade off one issue against another. From a philosophical perspective, he would like to approach this issue relationally, i.e., to mediate it so that all sides understand one another, and use their power to come to a mutual solution. But he wonders whether such an approach is viable when the political structure is set up as a win-lose situation.

He concludes that one must choose a power strategy to meet the political situation. As a science teacher, he can attend to the power relationships in his classroom and his school by using his authority, credibility, and expertise to construct "power with" relationships. The system may be small enough and the actors may have enough trust, that he can exercise power according to his values, i.e., as a means to transform a controversial situation into learning experiences, based on mutual gain and respect. But as a concerned citizen, he is not sure whether there is any way he can intervene in the airport dispute without appearing like a circle in a square. In this case, he can choose to either abstain, or to play the "power over" advocacy game.

The power flow diagrams have several useful instructional applications, especially when they are the subject of collective interpretations. As a small group of people view one another's charts, they are exposed to a variety of controversial situations and explore diverse perceptions of how power flows. They consider multiple strategic alternatives and solutions. Often a person is stuck in his or her perception of a political situation, whereas other people can offer different interpretations, novel intervention approaches that could have a significant impact. For example, one of my students created a power chart depicting what she perceived as an intractable conflict between the graduate students and the administration. Her colleagues suggested that her interpretation was biased by her experiences at other schools and perhaps by issues of authority. In fact, their observation was that this administration was flexible and open-minded, and quite willing to engage in a "power with" relationship. Where she saw confrontation, they perceived cooperation. Hence the power flow charts elucidate and amplify one's perception of power, the range of collective interpretations, and in some cases, the prospects for imaginative solutions.

Everyday Life as a Laboratory of Political Identity

To further amplify the power flow analysis, I invite my students to consider how public controversies emerge in everyday life and to analyze their own behavior in those situations. People learn about their political identity by carefully observing how they approach power and the methods they choose to resolve conflicts—the extent to which they avoid, defuse, mediate, or even instigate controversies. It is one thing to observe power as a detached bystander; it is another to become personally involved. Often people have values about power (i.e., the importance of relational power), but when conflicts emerge, events move quickly and one's deepest impulses may take over. As an example of how impulses challenge values, I describe to my class a situation in which *my* anger and irritation temporarily subverted my lofty philosophical notions about how to deal with conflict. I offer this story as a means of opening a collective discussion about political identity, conflict, and personal awareness.

During a backpacking trip in the White Mountains of New Hampshire, after a long stretch of hiking, my family (two adults and two children) arrives at a remote backcountry campsite. We decide to settle in for several nights. Immediately, we enter a "territorial" state of mind. Knowing that we will be living in this place, we look for the "best" campsite, which to us means privacy, an interesting view, and the appropriate intangibles. It is a wonderful feeling to set up the tents, organize our things, and settle into this wilderness place. This will be our temporary home.

During mid-August, even the most remote New Hampshire mountain will have some visitors, and an Appalachian Mountain Club "backcountry" campsite is sure to be filled by sundown. Every camper has a neighbor and every campground is a circumstantial community. Yet this is New England where good fences make good neighbors, so one assumes that backpackers will respect one another's privacy. People make the effort to trek into the wilderness because they wish to enjoy an outdoor challenge, but they also want a direct experience in nature, and appreciate simplicity and serenity.

After a full day at "our" spot, we are amply settled and relaxed. Upon returning from a day hike, we notice that there are some new neighbors. Somehow they managed to tote a boombox in with them and they are blasting heavy metal rock. In addition, the two men are speaking abusively (and loudly) to their female companion. This is all within earshot. It is impossible to avoid what is happening, although that is clearly our preference. Perhaps, we decide, we will ignore them and they will go away, or at least quiet down. But after several hours, our patience wears thin, and we have visions of a late-night heavy metal party. To be blunt, we become very pissed off.

Internally, I am fighting a battle. I know that I should probably just go over and talk to them, play the diplomat, and ask them if they could cut down on the noise. But instead, I conjure up awful stereotypes of heavy metal, redneck, sexist, uneducated creeps—my enemies, violating the sanctity of the wilderness, discourteous to their neighbors, disrespectful of the landscape, and generally evil. My vision of myself as a balanced, understanding, experienced controversial issues educator finally evaporates when I shout at the top of my lungs: "Would you turn that music off, I can't hear myself think!"

They lower the radio, but the war has begun. Our neighbors loudly discuss my request, presumably aware that we can hear their comments. They project their own stereotypes onto us, deciding who we are and what we stand for. "Folks who give their $100 to the Appalachian Mountain Club and assume they can buy their peace and quiet—some guy who thinks he's real important and is trying to get away from it all." Not that far from the truth, I think to myself. But this only increases my irritation.

Later we walk to a common area in the campground. Other backpackers concur with us, agreeing that the noise is intrusive. In effect, my family is on the front line for the entire community of campers. They suggest we speak to the caretaker, but that feels too much like calling in the police. We only want to use the legitimate authority of the caretaker if we can't solve the problem by ourselves.

We return to our campsite and sit several yards away from our tents. Our neighbors inspect our site, not knowing we are just a short distance way. They make disparaging remarks about us, poking fun at our tents, joking about the way the campsite is organized.

Something compels me to rush over to the evil campsite. I am not sure what my motivations are, and I find myself surprised to be there. My wife and children have no idea what I am up to and are upset that I rushed over without consulting them. I am in the enemy camp. Beer cans are scattered about. I wonder how they managed to drag up those six-packs. I quickly scan the campsite and I realize that the diplomat in me is taking over. I introduce myself, explaining that I am not there to hassle them, but since we are neighbors, we should get to know one another. After all, we are both in the mountains to enjoy ourselves.

After the uncomfortable introductions, the most belligerent of the three tells me that he lugged the boombox all the way up the mountain and he intends to use it. He says they are tired and they'll be out of our way real soon. I translate this as get the hell out of here, we won't bother you anymore. After a few uneasy moments, I thank them, wish them well, and return to my campsite. They turn off their radio and the incident is over.

When I first told this story to my class and to my colleagues, I felt slightly embarrassed, not because of my behavior, but because I wasn't entirely sure that the story was a meaningful political experience.

Perhaps the incident was trivial. Nothing really important was at stake. How can I compare the politics of this incident to the "serious" politics of civil wars, global poverty, and important elections? Who cares about the community politics of backcountry campsites?

Yet everyone I talked to reacted to the story by wanting to talk about similar things that had happened to them, wanting to compare responses. Many of my students said that these are the sorts of things that happen in life, and they are never sure exactly what to do. It opened a gold mine of conflict resolution storytelling and interpretation. My class assured me that the incident was vital because I attached deep moral and symbolic significance to the outcome. From our collective interpretation of the story I learned the following things.

I watched myself become angry, develop stereotypes, become stereotyped, and take a strong position based on my values and expectations. These values were attached to a moral position of right and wrong; therefore I cared about the outcome from several perspectives. First, I wanted my beliefs to be respected, that is, wilderness campsites should be places to experience peace and solitude. Second, I wanted to resolve the conflict based on a mutual understanding of each position. After all, that is what I "theoretically" espouse.

The campsite controversy served as an instructional metaphor for more serious political confrontations. Different parties have contrasting moral perspectives on issues of common concern. Stereotypes quickly emerge. Public issues may involve many more people, and the lines of power may be considerably more complex. But this small incident was important to me. I want to overcome my emotional impulse to exercise power for private interest. I place a high stake on achieving an outcome that is mutual and based on face-to-face discussion. I watched my temper subvert this prospect. I watched my moral judgments fuel my temper. This is a common loop in both private and public confrontations. It is a loop that often leads to suffering and violence.

But as people become more aware of how they behave in these situations, they can more effectively contribute to a community process for solving political problems, however they may occur. When I shouted to our neighbors and asked them to turn down the radio, I acted spontaneously. It was as if I was suddenly overcome by a wave of energy that forcced me to act. It was not deliberate. I acted impulsively, out of habit, out of emotion. When I walked to their campsite to initiate a discussion, I also acted spontaneously, but I was aware that I was acting from a moral center. I was acting within my values, knowing that I could not predict or control the outcome. I could try to understand their point of view, but also stand up for my convictions.

This became a broader political problem when other campers supported my position. It turned out that my family was on the front line because of our proximity to the noise. We had the most at stake. Certainly we were buttressed by the other campers who lent moral support to our cause. Hence we gained more power and confidence knowing we were acting on behalf of others as well. These situations are much more difficult when you are in the minority and there is very little public support for your action. It is important to experience the minority and majority perspective, because tomorrow we may be in the minority. Someone else may be on the front line.

My children observed my behavior very carefully. After all, I was acting as a role model, contributing to their political autobiography! I asked them what they thought of the situation. They told me they were scared when I went to the campsite. I told them that I was partly satisfied with my behavior in the situation. I was disappointed that I acted on my own without consulting the rest of the family. I described my anger, how my anger got fueled, how it prevented me from doing what I should have done in the first place, which was to go over to the campsite and make friendly contact. I wanted to convey, at the very least, that it is important to carefully observe your behavior in such situations and to act according to a moral center. I wanted them to know that this incident was a microcosm of larger and more complex political questions.

It is illustrative to observe how conflicts emerge among strangers in situations that are not ostensibly political. In circumstantial communities there are inevitable rules of common courtesy that are derived from implicit social contracts. Nevertheless, the implicit quality of these contracts implies that common courtesy is imprecise. Some may yield when others will take. Some may proceed when others will retreat. All may have different preconceptions of what it means to be a good neighbor. There are many different contexts in which people may be neighbors and wherever there are neighbors people jointly inhabit a place. There is a parallel between situations of common courtesy and one's treatment of the ecological commons. People can learn how to participate in decisions about the commons by understanding their approach to power in situations of common courtesy, the circumstances of everyday life.

The Power of Conversation

An extremely effective way to understand how power permeates everyday relationships is to explore how the use and interpretation of language provides people with access to political power. Robin Tolmach Lakoff in *Talking Power* examines the politics of language, noting that from "the most intimate tête-à-tête," to a "speech aimed at millions," the techniques of language are closely related.

> What is the relation between the subtle exercise of political strategies in friendly conversation, the more obvious clout special language gives the doctor or lawyer, and the persuasive blandishments of the political or commercial great communicator? All language is political; and we all are, or had better become politicians. [4]

Lakoff's book is about the relationship between language, politics, and power. What makes it so relevant for this discussion is that she is particularly interested in how these relationships permeate both everyday discussions and the sophisticated arena of international politics. For example, she distinguishes between macropolitics and micropolitics. *Macropolitics* refers to the "development and use of strategies that create and enhance power differences among individuals,"[5] such as struggles for power between groups, nations, institutions, etc. *Micropolitics* deals with strategies of group management and personal interaction, especially how relationships are established, who makes decisions within a group, and how direct the group communication will be. By looking carefully at how language is used in these circumstances, one gains insight about the use of power. Language can be used to divide or unite people. It can create change or it can reinforce the status quo.

For Lakoff, "our every interaction is political, whether we intend it to be or not; everything we do in the course of a day communicates our relative power, our desire for a particular sort of connection, or identification of the other as one who needs something from us."[6] The language of persuasion is intrinsic to many interpersonal exchanges. This is one way to learn about politics and power, by studying the language of everyday dialogue. Because language is so abstract and is something that is typically taken for granted, these connections can become explicit.

> Only metaphorically does language strike us or move us, and it changes us only in indirect ways. But as politics brings the brute reality of power into the

> sphere of the human mind and heart, language is the means of that transformation. Language drives politics and determines the success of political machinations. Language is the initiator and interpreter of power relations. Politics is language.[7]

If language is the "initiator and interpreter of power relations," then listening and speaking are the fundamental building blocks of political interaction. It is through ordinary discussion that ideas, opinions, and interests are expressed. Thus the dialogue of everyday life reflects a fundamental form of political expression. Speech interactions represent a complicated symbolic form of communication through which language is linked to consciousness. Throughout the day there are countless opportunities for political dialogue. These interactions may be mundane, but they frequently become extremely complex, having the potential to engender understanding or confusion, harmony or conflict.[8]

Sometimes conflicts emerge spontaneously, triggered by poor speaking or poor listening, resulting in a chain of misunderstandings, leading a conversation down a confusing path of distortions and allegations. This may be intentional, but typically it is not. Rather, stereotypes and preconceptions predispose people to hear what they want so that it conforms to whatever version of the truth they think they're carrying. This process occurs both in private conversations, which are ostensibly nonpolitical, and in public discussions addressing contemporary issues.

Listening and speaking skills are fundamental to mutual, participatory, political interactions. People speak to make themselves understood. They listen to understand others. Yet the complexity of speech interactions, both the contextual and perceptual dynamics which inform discourse, make listening and speaking, having a good conversation, an extraordinary challenge. People often speak and listen automatically, habitually, and without reflection, not knowing where their words come from, where they go, how they lead to new words, and what kind of impact they will have. It is true that people exercise censorship between what they think and what they say, but often they don't say exactly what they mean, or what they say is not interpreted the way it was meant, or they don't get their ideas across as well as they'd like. Interactive clarity is best achieved when people think, speak, and listen reflectively linking intention to language, awareness to thought, and mindfulness to communication.

Benjamin Barber claims that the heart of participatory democracy is talk. He describes the "mutualistic art" of listening, suggesting that good listening means putting oneself in another person's shoes, looking for common ground, seeking a true understanding of the other position, allowing people to generate compassion and establish a connection of heart and mind.

> Listening is a mutualistic art that by its very practice enhances equality. The empathetic listener becomes more like his interlocutor as the two bridge the differences between them by conversation and mutual understanding. Indeed, one measure of healthy political talk is the amount of silence it permits and encourages, for silence is the precious medium in which reflection is nurtured and empathy can grow. Without it, there is only the babble of raucous interests and insistent rights vying for the deaf ears of impatient adversaries.[9]

Barber shows how political talk should have both affective and cognitive modes, allowing people to explore the full range of their feelings, impressions, and intuition about an issue. Talk links thought to action, by allowing people to creatively invent new ideas, and establishing alternative visions for their common interests. During a good conversation, ideas become real and tangible. Thus, speaking and listening are the core of political communication.

Controversial issues involve a complex web of power relationships, intentional manipulations, and ideological predispositions. The concerned citizen can cut through these walls of controversy by learning how to construct vivid, sincere, and clear dialogues in which the common interests of all parties are served. This task is fundamental to political learning, the ability to use everyday language as a way to participate in the great power struggles of macro- and micropolitics. Political identity work is practiced by learning how to have a good conversation.

The Power of Controversy

But how much influence can people have on the *quality* of public conversations about national or international issues? This section explores some of the dilemmas that emerge when controversial issues are dominated by "power over" approaches. When public dialogue is influenced by manipulative, propagandistic language, should the environmental activist or concerned citizen endorse similar tactics to influence a decision? Or is it more important to elevate the moral integrity of the conversation, regardless of the policy outcome? These difficult questions are more than strategic alternatives—they are at the heart of ecological citizenship, challenging people to consider the quality *and* morality of their public involvement.

In the 1980s, during the height of the nuclear power controversy, various political advocacy groups ran a series of advertising campaigns to convince the public of the virtues of their cause. The Committee for Energy Awareness, which was heavily funded by the utility industry, ran an extensive series of magazine and television advertisements, proclaiming the safety, necessity, and logic of nuclear power as an energy source. Their magazine advertisements were "fact and figure" ads, that is, they tried to build a reasonable, unemotional case, based on a particular interpretation of American energy policy. The advertisements looked like *Scientific American* articles; they were well constructed, visually appealing, and reflected an image of utter credibility. Their television ads were much more emotional, using various symbols and backdrops to declare the urgency of energy independence and economic growth—linking those themes to patriotism, family values, and nuclear power. Both the television and magazine ads tried to convey an image of open-mindedness, citing the importance of diverse energy resources, but clearly they supported nuclear power.[10]

As a response to these ads, an environmental advocacy group (The Safe Energy Council) that supported alternative energy sources and emphasized energy conservation developed a series of advertisements that were strongly antinuclear. This group did not have the same funding resources as the Committee for Energy Awareness so they had to target their ads very carefully. They designed an advertising campaign based on the Three Mile Island accident. The ads were very symbolic and deeply emotional, clearly using fear and manipulation to influence people about the dangers of nuclear power. It is not a stretch to claim that these ads were propagandistic.

Both groups used propaganda and manipulation in their advertisements. Political advertising has become a normative component of any strategic political campaign and we need look no further than presidential elections to bear this out. These tactics are commonplace in American politics, including environmental issues. The language of persuasion is brought heavily to bear on lobbying and advocacy.

The ethical ramifications of this situation are salient and compelling. If you believe fervently in the dangers of nuclear power, would you sanction the use of propagandistic advertisements if you thought they would significantly influence public opinion? I decided to pose this question to my class. Most of my students are opposed to nuclear power and are strong proponents of energy conservation and alternative energy sources. I showed them the video of the "manipulative" advertisement asking them whether they would authorize the use of the ad. A heated discussion followed regarding the ethics of political advertising, the use of propaganda, means versus ends, urgency versus patience, and "power over" versus "power with."

Because my students typically have a great deal of professional work experience, many have been in positions in which their organizations faced similar questions. I ask them what decision they would make as the director of such an organization (some are currently directors) if they were faced with an urgent environmental question and were opposed by a well-financed advocacy group that would use whatever manipulative tools it had at its disposal. Would they sanction the use of political methods that involve propaganda and devious manipulation?

This is a confounding issue for many environmentalists and concerned citizens. My students and most environmental professionals I speak with are split on this question, not only among themselves, but often within themselves. They understand that to run an effective advocacy campaign, they must appeal to people emotionally as well as scientifically, and that in certain political circumstances, this may tempt them to sanction propagandistic tactics. Yet they are also uncomfortable with this approach because it seems ethically inappropriate. In addition, many environmental professionals have dual roles. Sometimes they wear the hat of the advocate, at other times they wear the hat of the educator. However, education and advocacy are not the same thing.

Education about environmental issues is often taken up by groups who have a specific advocacy agenda. Sometimes they honestly assume that they will be able to subdue their opinions and present clearly balanced arguments representing several perspectives on an issue. However, some educators are really advocates for a particular perspective, although they are masquerading behind a facade of "values-free" teaching. This is a common problem in environmental education. Advocacy groups (corporations or citizen action groups) often produce educational materials that purport to describe the facts surrounding an issue objectively. But sometimes these materials are merely forms of propaganda designed to convince the user of a specific point of view. These materials may appear as "technical information" written by "experts" in the field with the purpose of providing the reader with what is supposed to be scientifically sound information. Environmental issues, similar to most controversial issues, are replete with hidden biases that distort supposedly objective statements, propagandistic statements that are misrepresented as fact, and statements of deep ideological conviction that appear as fundamental truth.

In any controversial issue, people are tempted to present a biased point of view in order to make their point. Sometimes, despite their best intentions, people unknowingly manipulate information by their arrangement of the sequence of presentation, selection of materials, use of subtle body language, or admission of selected evidence. Teachers, administrators, environmental advocates, businesspersons, concerned citizens, whatever their role, people are faced with situations in which they have to decide whether or not they will manipulate a group for strategic purposes.

This is also the domain of political identity work. By using propaganda and manipulation to achieve a strategic end, one implicitly endorses a "power over" approach to politics. By upholding the standards of a mutual, relational, collaborative conversation, one endorses a "power with" approach. Yet political situations are rarely so cut-and-dried, and often the urgency of the issue, or the context of the debate, will influence how one uses power. In these situations, one may win the debate, but sacrifice the ideals of a "good" public conversation. Is any situation worth this sacrifice, or are there alternative means for accomplishing one's goals?

A person's view of human behavior will significantly influence his or her political identity. For example, if you think that humans act only in their self-interest and require a strong government to harness their idiosyncratic impulses, then you may suggest coercive measures as the most effective way to solve environmental problems. If you believe that humans are likely to act in the common good and are relatively altruistic, especially in the absence of a coercive and restrictive state, you may have more faith in consensus-building and negotiation.

I ask my class to consider how their view of power influences their approach to the following controversial dilemma. When the circumstances are dire—let us say your interpretation of scientific evidence compels you to believe that some type of environmental catastrophe is inevitable—what is the extent to which you would implement coercive policy to remedy the critical problem? There is a copious literature, especially among political scientists, that addresses this question. It emerged during the 1970s when the first limits-to-growth studies were released, prompting a renewal of neo-Malthusian doom and speculation. At least four writers from different disciplines (Paul Ehrlich, Garrett Hardin, Robert Heilbroner, and William Ophuls) were extremely pessimistic about the prospects of liberal democratic institutions being able to handle the environmental crisis. They all urged various Hobbesian solutions.[11]

More recently, population biologist Garrett Hardin, who wrote about the tragedy of the commons in the 1970s, has once again proclaimed the importance of strong coercive measures. In his recent book, *Living within Limits,* Hardin concludes that governmental coercion of some sort is needed to restrict human population. The role of government coercion as an instrument of policy has been openly debated in the annals of environmental texts as well as in the halls of public policy. It is crucial to face this question openly and honestly so that environmentalists can decide how committed they are to the democratic process. This is ultimately an ethical question and it is something all environmentalists face whether they work for a conservation commission or a federal policy institution.

I ask my students whether they would approve of an environmentaly oriented decision-making elite to act as a Supreme Court of public policy and whether they would mandate that elite to enforce human population limits through reproductive restrictions. Or, would they prefer to rely on moral persuasion and environmental education? People find themselves split by these controversial and difficult questions. Some believe that such elites inevitably fall prey to their own power. They wonder what would make an "environmental" elite any different from other decision-making elites. They claim that significant environmental reform is impossible unless it is based on a change in the way people think—you can't cultivate environmental values through coercion. Others assume that there is no alternative. Powerful decision-making elites are a political reality. Why not construct an ecologically oriented elite to implement and strongly enforce environmental policy? Isn't this what politics is all about? For them environmental pollution is so urgent and so threatening that it requires the strongest possible action. But what do we mean by "strong action?" And what guarantee does anyone have that coercion doesn't breed more coercion?

These dilemmas exist on every level of political decision making. As a faculty member in an environmental studies program with 350 graduate students, in an institution that prides itself on social change and democratic process, I am constantly watching administrators and faculty wrestle with the question of the extent to which they should include either students or support staff in important decisions. On the one hand, people with power say they value the democratic process, yet they also recognize the importance of administrative efficiency. I often watch people (including myself) squirm out of the dilemma by offering token forms of inclusion. I have seen educational institutions become paralyzed because they became too inclusive and couldn't develop an efficient decision-making process. I mention this merely to express how issues of power and restraint, inclusion and exclusion, hierarchy and decentralization permeate so many aspects of political identity. These are profound dilemmas that will not be solved with the certainty of ideological rhetoric. Are we willing to jeopardize democratic process when we believe strongly in a cause?

The quality of political participation is as important as the urgency of ecological reform. If the structures and processes of normative politics seem unresponsive or exclusive, then it is our responsibility to either change those structures or construct new ones. For example, within the last 10 years the environmental dispute resolution process has been used as a policy alternative when normal political channels were ineffective. This approach is increasingly used to solve difficult environmental problems regarding the commons. The field of international environmental politics has undergone rapid conceptual revision as activists, policymakers, and ordinary citizens explore new protocols and regimens of transboundary decision making. The widespread concern about global environmental change has spawned new coalitions and arrangements of grassroots activism. And even entrenched bureaucracies can change from within. The United States Department of Energy, in a stunning policy reversal, opened the "files" on the nuclear experiments of the 1950s. This was a milestone both for democratic process and ecological reform.[12]

Surely there are circumstances in which people lose their patience and begin to lose faith in democratic process. At that point, they must remember that the purpose of politics is to find out what people have in common, discovering the ground (literally) of affiliation, mutuality, and common interest. There are methods that "stuck" groups can use to make their decision-making processes more efficient. They can do their best to compromise, but also be willing to submit to mediation or arbitration when compromise breaks down. Sometimes a majority decision is required; sometimes a group must trust the judgment of its leaders. The majority is not always right, nor is leadership always wise; that is why we must have accountability and evaluation. The minority may not get what they want, but they should be guaranteed full expression and respect.

Nor should people temper their advocacy when they strongly believe in something. It is important to understand diverse perspectives, but sometimes you have to take a stand. Controversy is intrinsic to environmental issues. It is not easy to carry the burden of controversy. Although democratic society cherishes the virtues of pluralism, it is hard for any individual to be controversial. Controversy begets attention, demonstrates difference, introduces new ideas, disrupts easy explanations, challenges values, and places us under careful public scrutiny. As a political identity checklist, I ask my students the following questions: Do you take a stand on issues that are important to you? Or do you hold your credibility hostage to unconventional or unpopular points of view? Do you have the political capacity to state your deepest opinions without alienating your neighbors or colleagues? Can you speak your mind, but also really listen to what other people have to say?[13]

As long as a person participates in decision making about the commons, he or she will espouse and encounter controversial views of the world. These views will challenge other people, encouraging them to think about how they live their lives. For the environmentalist and concerned citizen, there is no escape from controversy. It is through the convergence and comparison of diverse perspectives that politics becomes a learning process, the basis of personal and political transformation. When this occurs, people traverse the "hidden valley of shared values" and find the ecological, political, and moral common ground of their humanity.

Toward Ecological Citizenship

The educator encourages ecological citizenship by exploring how ecological and political identity comprise intersecting sets of wider identification. Consider the following example as an illustration of this process.

Suppose that for breakfast I toast a bagel. When I use the toaster I am connected to an electric energy grid. The energy may be coming from the Seabrook nuclear power plant or from Hydro-Quebec; both controversial energy sources. If I carefully trace the energy to its source, I unravel a complex web of ecological and political relationships. Seabrook is the source of unending controversy, symbolic of many of the questions surrounding the viability of nuclear power—issues of scale and decentralization vs. corporate power, issues about uranium mining and waste disposal, questions of public input into utility decision making. Hydro-Quebec is involved in massive controversial development schemes, which have a profound impact on the political landscape, providing energy and economic growth, but threatening the cultural and ecological integrity of indigenous peoples. Whether I choose to toast a bagel, or eat out at McDonald's, or collect wild berries in the forest, my actions are much more complex than they initially appear.[14]

How does breakfast integrate ecological and political identity? One can *observe* the wider circles of identification. The act of preparing, consuming, and disposing of breakfast is much more than a choice of cuisine. It can be linked to the world of contemporary events: the great flows of national and international power (global energy policy), the small details of community politics (do we really need a McDonald's in our neighborhood?), or the changing shape of power over space and time (who gets to eat breakfast?). It can be linked to interconnected ecosystems: the grand scope of ecology and evolution (where does wheat come from?), the news of the landscape (organically grown?), the daily life of an ecological place (is it a good year for wild berries?). Rarely does a person consider all of these dimensions when he or she sits down at the breakfast table. Through the cumulative observation of dozens of such ordinary life activities, the patterns of ecological and political interconnection begin to emerge. As an ecologically minded citizen, it is my responsibility to observe the wider impact of my daily habits.

Through the lens of ecological and political identity work, one can *interpret* the deeper meaning of breakfast. How do my breakfast habits correspond to my ecological and political values? By eating at McDonald's do I contribute to a way of life that runs counter to what I believe in? Interpretation connects what one observes to what one thinks, lending meaning to events and relationships by evaluating, analyzing, and critiquing one's thoughts and actions. There are multiple plausible interpretations of my breakfast experience, but I can choose to frame them from the perspective of ecological and political identity work. Breakfast becomes part of my life story. Consider the following interpretive path.

Whenever possible, I eat organic grains, produced by socially responsible companies, thus contributing to my personal health, the ecological health of the planet, and politically correct business. I interpret these actions as responsible ecological citizenship. Yet often I eat breakfast much too quickly, falling prey to the demands of a busy life in which I feel compelled to accomplish and achieve a great deal, orienting my conception of time around a logical, efficient sequence of measurable tasks and objectives. The culture of achievement supports this time management approach. As I eat, I look at the clock, to make sure I won't be late for an appointment. I recall Lewis Mumford's description of the clock as a means of disassociating time from organic events, allowing for the abstractions of time as money and money as power.[15] Am I trapped by the modern conception of industrial time? As an ecologically minded citizen, it is my responsibility to interpret the consequences of my actions, to make them coherent, to place them in a broader context, to consider their deep meaning, and to link them to my espoused values.

The crux of citizenship is to *participate* in discussions about the commons. To adhere to my values, I might decide to avoid McDonald's and eat at the local organic foods restaurant. But this is only a consumer choice. I participate in public issues when I call attention to the reasons I make the choice, and raise them in a larger community setting. There are dozens of reasons why McDonald's should or should not be built in the neighborhood. I act as an ecologically responsible citizen when I become publicly involved, joining with my neighbors in thinking through the decision, advocating my values, and listening to other positions. When I toast my bagel, I may feel limited by my consumer energy choices—it may be impractical to go off the electric energy grid. But I can participate in the political process of New England energy policy by raising public awareness, setting up public forums, or taking a strong advocacy role, whatever actions fit my political temperament.

Participation is the laboratory of ecological citizenship. Benjamin Barber, in *Strong Democracy*, describes the relationship between community and participation.

> Community grows out of participation and at the same time makes participation possible; civic activity educates individuals how to think publicly as citizens even as citizenship informs civic activity with the required sense of pub-

licness and justice. Politics becomes its own university, citizenship its own training ground, and participation its own tutor . . . to be a citizen is to participate in a certain conscious fashion that presumes awareness of and engagement with others. This consciousness alters attitudes and lends to participation that sense of the we I have associated with community. To participate is to create a community that governs itself, and to create a self-governing community is to participate. Indeed from the perspective of strong democracy, the two terms participation and community are aspects of one single mode of being: citizenship.[16]

Ecologically responsible citizenship involves both the content and process of politics—the environmental issues that constitute the commons, and the community building that emerges from public participation.[17] Indeed, this is the intersecting set of ecological and political identity, the merger of an ecological worldview and civic awareness, through the interpretive strength of personal reflection. Ecologically responsible citizenship becomes the arena of personal transformation as public life reflects the moral dimensions of everyday choices.

Vaclav Havel, in *Summer Meditations,* lends a moral and spiritual aspect to citizenship. In his role as statesman he tries to awaken the dormant goodwill in people because "goodwill longs to be recognized and cultivated."[18] People must be reminded and told that "it makes sense to behave decently or to help others, to place common interests above their own, to respect the elementary rules of human coexistence." For Havel, these are basic principles of political leadership. Because without this moral sense, there cannot be a moral state.

A moral and intellectual state cannot be established through a constitution, or through law, or through directives, but only through complex, long-term, and never-ending work involving education and self-education. What is needed is lively and responsible consideration of every political step, every decision; a constant stress on moral deliberation and judgment; continued self-examination and self-analysis; an endless rethinking of our priorities. It is not, in short, something we can declare or introduce. It is a way of going about things, and it demands the courage to breathe moral and spiritual motivation into everything, to seek the human dimension of all things.[19]

Both Barber and Havel refer to a profound mindfulness in relation to politics. Political identity is more than just the content of public issues, rather it is the moral deliberation that guides personal and collective action. It is a process of deep introspection, a means of self-discovery through one's identification with a larger whole: the common domain of human meaning.

The ecologically aware citizen takes responsibility for the place where he or she lives, understands the importance of making collective decisions regarding the commons, seeks to contribute to the common good, identifies with bioregions and ecosystems rather than obsolete nation-states or transnational corporations, considers the wider impact of his or her actions, is committed to mutual and collaborative community building, observes the flow of power in controversial issues, attends to the quality of interpersonal relationships in political discourse, and acts according to his or her convictions. The ecologically responsible citizen recognizes that he or she lives a life in nature, in conjunction with other people, in the common interest. Where does one practice this approach to life if not in the common domain?

5 Ecological Identity and Healing

A fourth-grade boy sits in his school library, searching the latest issue of *National Wildlife* for a report on ducks he is preparing for a wetlands education project. He discovers a grotesque picture of a duck covered with black oil. Horrified and intrigued, he quickly turns the page and finds several appealing illustrations of unspoiled wildlife. Later, he asks his teacher why the oil is on the duck, whether the duck will survive, and if the oil poses a threat to Stanley Brook, the wetland his class is exploring. His teacher isn't sure how to respond. She has organized the project around natural history and sensory awareness. Although the children talk about threats to the wetland, such disturbing images haven't emerged so vividly. She wonders about the psychological impact of the oil-covered duck, and what her responsibility is in helping the child deal with the emotional consequences.

A wetlands scientist spends weeks studying a wetland, preparing an environmental impact statement for her consulting firm. She lists all the reasons for protecting the habitat, citing its ecological diversity, the purifying role it serves for the watershed, as a habitat for endangered species, not to mention its aesthetic value. Although she lives 50 miles away, she develops a sense of attachment to the wetland. She cares about its future, desperately hopes it will remain protected. After several weeks of policy deliberation, a compromise is reached, but it includes a degree of development that she is convinced will have irreversible ecological consequences. From her perspective, the wetland is now doomed, and she feels that it is her responsibility to convey this message to the public.

A parent takes his children to visit the Boston Science Museum. They attend the Omni Theatre presentation on the tropical rain forest. Immersed in a wave of colors and sounds, they are bathed in a profusion of spectacular footage of insect and plant life. The image is interrupted by an abrupt discontinuity: a chain saw savagely destroying trees, forest fires visible from space, erosion and decline. The film includes a quick lesson in biodiversity and how the destruction of even a small area of the rain forest can have severe consequences for all the species that live there. Just as suddenly, the film ends, the lights come on, and he and his family leave the theater, ready to view the next attraction. The man wonders what emotional impact, if any, the film has had. Is it just another electronic spectacle, or do people carry this sense of loss with them?

An environmental educator describes a recent camping trip to Mohawk State Forest in western Massachusetts. After participating in an interpretive walk (the kind of work she usually does herself), in which she learns about the beauty and fragility of the local landscape, she strolls into the bathroom. On the door is a poster explaining what to do in the event of a nuclear accident at the Rowe nuclear power plant. She is left with a distinct ambivalence: should I love this land or not?

When I speak with environmental practitioners about the psychological impact of their work, responses such as these typically emerge. Whether they wear their professional hats as educators, managers, scientists, or activists, or if they respond as concerned citizens who observe the world around them, they are faced with a daunting, even overwhelming challenge: how can they simultaneously promote awareness, enjoyment, and sensitivity to the natural world, while pointing out all of the ominous threats to global ecosystems, local habitats, and human well-being? How can they at once be conveyors of wonder and harbingers of doom? In my experience, it is evident that people need support for their feelings and concerns about these questions, and they are deeply appreciative of discussions about whether and how to present such challenges to the general public.

The psychospiritual ramifications of internalizing global environmental change are crucial to ecological identity work, yet surprisingly few environmental practitioners consider this dynamic in their formal educational training, and rarely do their organizations discuss the topic. They are busy attending to the practical work of implementing urgent environmental reform, promoting an ecological worldview, or advocating their opinions. But there is an affective detritus, sifting through the cracks and settling on the bottom of a task-driven professional life—feelings of anxiety, despair, and grief juxtaposed with reverence, compassion, and wonder.

This is the striking emotional message conveyed through the symbols and texts of contemporary environmentalism. Whether reading the newspaper, watching the evening news, surfing the Internet, or observing the changes in one's habitat and neighborhood, we are all exposed to a daily barrage of ugly environmental images. This prompts a range of responses. Some of us become psychologically numb as we become accustomed to the litany of environmental bad news. Others experience fear and anger, are outraged by the social and environmental injustices that plague the planet. Sometimes people just tune out and ignore or avoid the negative images, rationalizing their inaction, practicing denial and apathy. The profound challenge is to learn to reflect on the emotional consequences of these perceptions and images, to use them both to motivate action and to look deeply within oneself. With the exception of the groundbreaking "despair and empowerment" work of Joanna Macy and some recent innovative approaches in ecospirituality (both described later in the chapter), environmentalists tend to underestimate the psychospiritual consequences of their work.

In many respects, environmental practitioners are involved in a healing profession, yet they don't necessarily think of their work in such terms. When an environmental technician cleans a toxic waste site, when an environmental educator explains the causes and consequences of an oil spill, or when an environmental scientist restores a wetland, these are all examples of personal and community healing. The restored wetland promotes community well-being, protecting the habitat from environmental pollution and threats to its residents. The schoolroom discussion of an oil spill allows students and teacher to consider the ecological devastation, express their fears and concerns, construct policy and lifestyle alternatives that will influence the life choices that a person may have. Environmental issues cannot be solved with merely a technical fix—they require an understanding of the ecosystem, of the human condition, and of the psychospiritual wholeness that is necessary to one's experience and perception of nature. From this perspective, ecological identity work has a profound healing agenda: restoring ecosystem health, community well-being, and personal happiness.

This chapter explores the relationship between ecological identity and healing from three perspectives. First, it considers the images of hope and despair that pervade environmental work, suggesting that environmentalists have a responsibility not only to advocate for ecological reform and to inculcate appreciation of the natural world but also to provide support for the anxiety that accompanies the perception of cultural upheaval and wounded ecosystems. Second, it considers the psychological implications of acknowledging anger and despair. How can blame and guilt be transformed into responsibility and action? Third, it describes some of the conditions of organizational stress that prevent environmental practitioners from confronting the psychospiritual implications of their work. How can the environmental practitioner bring a reflective perspective to professional life and what are the educational dimensions of such a task?

Conveyors of Wonder/Harbingers of Doom

During the 1980 presidential campaign, Ronald Reagan and Jimmy Carter debated the priorities of the American people. Through their advertisements, speeches, body language, and platforms, they presented starkly different perspectives on the future. Carter was the stern moralist. He preached the necessity of limits, the importance of sacrifice, and questioned the economic and spiritual virtue of unlimited affluence. Reagan, on the other hand, was buoyantly optimistic. He extolled the American dream, and couched it in crassly material terms. He promised to deliver the joy and bounty of unlimited economic growth—enough for everybody, no sacrifice necessary, a future of unbounded glory and wealth. Of course, Reagan won the election in a landslide, in part because of his optimistic message. Carter's vision was just too bleak.

Environmentalists carry some of the same baggage as Jimmy Carter. Often, they are perceived as the messengers of doom, Cassandras of contemporary life, conveying warnings and admonitions, spreading bad news, reminding people of the dire consequences of the profligate lifestyle. Modern environmentalism involves an apocalyptic dimension, portraying nasty scenarios and metaphors: silent spring, population bomb, acid rain, toxic waste, resource scarcity, nuclear winter, ozone hole, global warming, and species extinction. People just don't like to hear about such things and when they do, they are not at all sure how to respond.

Environmentalism has many constructive metaphors and scenarios as well—reinhabitation, safe energy, organic farming, recycling, appropriate technology, small is beautiful, ecological restoration, and sustainability—suggesting dozens of practical alternatives, from everyday life choices to the complex machinations of public policy. But at the bottom of all of these suggestions is the implication that something is wrong with the way people live their lives. Whether it's a soft nudge to consider recycling, a hard push to rethink the foundations of everyday habits, or a harsh critique of industrial civilization, the theme is the same: the planet is in trouble and everyone has a responsibility to do something about it.

Many environmental professionals work with the general public, attempting to cultivate environmental awareness, which includes not only understanding the threats to nature but developing sensitivity, joy, and wonder for the diversity of life. Indeed, this "awareness" includes a scientific, aesthetic, and even spiritual appreciation. For example, when an environmental educator develops a curriculum or runs a public program designed to promote sensory awareness of nature, he or she is involved in a form of ecological identity work, promoting identification with the landscape in order to cultivate deeper appreciation. The educator usually has multiple objectives: teaching natural history and field ecology to enhance ecological understanding; stimulating a sense of wonder and discovery to enhance moral and spiritual growth; presenting practical, everyday choices to promote ecologically sound lifestyles; and pointing out the threats of environmental pollution to galvanize public action and concern.

These objectives are often framed as a series of moral choices. It is likely that the educator, similar to thousands of other environmental practitioners, is either preaching a moral homily, or rousing people to convert to his or her ecological worldview. Environmental practitioners become, accordingly, stereotyped, and are sometimes perceived as moralists.

One element of reflective practice (described in more detail later) is the ability to recognize how others perceive your work. This is helpful in several ways—it cultivates understanding of the needs and perspectives of a client or audience, it develops personal awareness of how values and ideals are communicated in a professional context, and it reveals the stereotypes that may frame your actions. When I work with environmental practitioners, as a first step in considering reflective practice, I ask them to assess how their environmental persona is perceived, either in the workplace or in everyday life. Numerous examples of the "environmental stereotype" emerge, confirming that most environmental practitioners are viewed as espousing an ecological worldview, and this effects how they are perceived by clients, friends, and neighbors. There is baggage that accompanies this perception.

Larry, an environmental administrator, describes how he is perceived by a neighbor:

When I am out walking in the woods, I sometimes encounter a neighbor who is a logger. We form a common bond based on the fact that we typically see each other outdoors and that we rarely run into anyone else out there. We may be in the woods for somewhat different purposes, but clearly we both enjoy the outdoors. We get into long conversations. He typically makes some comment about how he has done something that makes environmental sense, as if I am judging him, or to prove that we have something in common. I find that I want him to know that I'm just a normal person. Nevertheless, it's clear that I carry a stereotype, and it is very much of my own making.

My neighbor was responsible for logging a piece of land on which I regularly roam. Although he did a reasonable job in terms of erosion and sustainability, the landscape was scarred, and in some places became impossible to traverse. Yet he was proud of the work he had done, feeling that given the topography and economic factors, he paid attention to ecological concerns. From my perspective, he dramatically transformed the landscape in a way that was aesthetically degrading and disturbed the spirit of the land. I couldn't tell him this directly because I wanted to encourage his new acknowledgment of environmental responsibility, without sounding like a Druid. I don't blame him for what he did, but I haven't reconciled my feelings about the way the landscape was changed. And I haven't figured out how to talk to him about these feelings.

Many environmental practitioners describe how they are treated by friends and neighbors. Consider this typical comment, in this case from Jessica, an environmental educator:

When I visit a friend, she inevitably feels like she has to change her house around before I visit, as if I am the environmental police. She spends too much time apologizing for her inappropriate environmental behaviors. Why does she assume that I am always judging her? Is it something about my behavior, the behavior of environmentalists generally, or is it her way of telling me that she would like to live more responsibly? This type of thing happens to me all the time. When I tell people I am an environmental educator, even strangers, they immediately assume that I stand for certain things. They prejudge me, just as, perhaps, many environmentalists prejudge them.

Marie, a social worker, sheepishly admits that she has taken actions that contribute to the stereotype:

I had just finished taking a hike at a nearby state forest. As I arrived at the parking lot I noticed some people dumping the contents of their car ashtray into the woods. I was incensed. The people drove off so I decided to follow them. I did this for over half an hour. Finally the outlaws parked their car. I pulled out of my car and confronted them. They were a "sweet" elderly couple and they were terrified of me. Why was I following them? While I was explaining to them why it was so negligent to dump their ashtray in the woods, I realized that they had no awareness that they were doing anything wrong. They had a completely different value system. The couple thanked me for educating them about their deed. I realized that I wasn't really confronting them as much as I was confronting myself, acting on my own need to deal with anger and dismay.

Environmentalists are viewed as having a particular set of opinions about the right way to live. They have to simultaneously substantiate their observations of environmental decline and demonstrate the viability of practical, ecologically sound alternatives. And if they take these notions seriously, they undertake an enormous responsibility. In the examples above, the environmental administrator is obligated to confront his feelings about his neighbor's logging project and to somehow engage his neighbor in a deeper conversation; the environmental educator realizes that she carries more than her own persona, but she conveys the message of contemporary environmentalism; the social worker understands that in her moral fervor, she has neglected processing her own deep feelings about destructive environmental acts.

Recently I was browsing through the children's section at a local bookstore. I was particularly interested in the nature and environment shelves. There were many volumes, profusely illustrated, explaining all aspects of the environmental crisis. These were compelling books, of interest to adults and children alike, designed to cultivate environmental awareness, including nature appreciation and an understanding of the threats to nature, providing a great deal of useful and accessible information. Yet few of these books seemed to acknowledge the feelings that a child has when confronted with the bleak outlook contained in their messages. I saw books on endangered species, endangered places, global environmental change, and environmental pollution, that didn't offer the child or parent any advice on how to cope with the material. What is a child or an adult to think when he or she gazes through these books, reading about dire environmental threats, but is left on his or her own to contemplate the emotional ramifications?

Or consider the recent book by the Earth Works Group, *50 Simple Things You Can Do to Save the Earth*. The first section briefly summarizes the severity of the environmental crisis. The 50 things to do are organized according to "simple things," "it takes some effort," and "for the committed." The book is filled with excellent suggestions, accessibly presented and provides the interested reader with dozens of constructive lifestyle alternatives. But there is a glaring absence of thoughts, suggestions, or even bibliographical references to the psychospiritual implications of its content.

Anyone who has even marginal contact with the daily television news is aware of the various prognostications of ecological doom. Often, it is the role of the environmental professional to create public awareness of this prospect. Yet it is very difficult for people to assimilate this information. The concerned citizen cannot escape the loss and suffering that accompany ecological deterioration. However people choose to understand the magnitude of this problem, through emotional pain, or sheer deductive analysis, the psychological turmoil of global environmental change is inevitable. Many environmentalists are unwilling to confront the necessary despair work because they just don't know how to do so. They carry planetary distress as a nagging, chronic pain, never really plunging into its psychospiritual meaning, often responding with an overriding anger, and using this anger as a motivation to take incessant action. Sometimes it causes them to burn out on environmental work.

James Thornton, an environmental attorney and activist, recently interviewed fifty environmentalists in an effort to determine, among other things, the role of anger and cynicism in the activist's work, whether there was any sense to having a support community in times of stress, and whether the activist's institutional setting fostered personal growth. In his summary observations, he writes:

> Many activists believe that the basis of their work is anger. It was often stated in the interviews that an activist needed his or her anger to be effective. While many understood that there was a connection between using anger as the basis of action and burnout, there was an often stated reluctance to give up anger. This reluctance held a fear: the activist didn't know any other basis of effective action than anger, and was worried that if anger was not present, there could be no action. . . . Nevertheless, there was a universal curiosity about and interest in some form of personal work. It was widely felt that some form of training could offer the possibility of more effectively working with stress, and coming to a deeper understanding of root issues.[1]

Thornton is concerned that for many people, there is no way through the anger and despair that accompanies environmental work. He suggests that this may be one reason why "the environmental movement has tended to preach in strident, negative tones." Thornton proposes that environmentalists engage in what he calls "wisdom training," which combines intimate experiences with the natural world and deep self-reflection in order to liberate the anger and despair, and use it creatively to develop "radical confidence."[2]

Joanna Macy provides numerous educational approaches to understanding the experience of ecological despair and its relationship to community healing. She describes how "a dread of what is happening to our future stays on the fringes of awareness, too deep to name and too fearsome to face." She emphasizes how important it is that these feelings come to the surface:

> Despair cannot be banished by injections of optimism or sermons on 'positive thinking.' Like grief, it must be acknowledged and worked through. This means it must be named and validated as a healthy, normal human response to the situation we find ourselves in. Faced and experienced, its power can be used, as the frozen defenses of the psyche thaw and new energies are released. Something analogous to grief work is in order. "Despair work" is different from griefwork in that its aim is not acceptance of loss—indeed, the "loss" has not yet occurred and is hardly to be "accepted." But it is similar in the dynamics unleashed by the willingness to acknowledge, feel and express inner pain. From my own work and that of others, I know that we can come to terms with apocalyptic anxieties in ways that are integrative and liberating, opening awareness not only to planetary distress, but also to the hope inherent in our own capacity to change.[3]

Macy explains how when people open themselves to the pain and suffering of despair, they encounter some "psychic disarray," suggesting that such a process is constructive if handled properly. She refers to "positive disintegration" in which people enter spiritual turbulence, watching their defenses and walls break down, but allowing them to consider "new and original approaches to reality." To experience the suffering of planetary distress is to "permit ourselves to feel." By acknowledging these feelings, people summon deep collective energies, create communities of empowerment, become able to discuss their feelings, work together to form reservoirs of strength and vision, and take responsibility for transforming their fear and suffering into commitment and action. Out of this community springs hope and vitality, the ability to work creatively to overcome the sleep of denial and forgetfulness.

John C. Elder, in his article "The Turtle in the Leaves," explores the redemptive quality of grief, and how perceptions of loss and despair are intricately tied to wholeness, leading ultimately to hope:

> Grief, as each of us has the chance to discover for ourselves, may be distilled to hope, but only if we can bring ourselves to affirm both the inevitability of our loss and the beautiful wholeness of which it constitutes a part. Such affirmation depends in turn on familiarity, and identification, with the larger encompassing cycles—through which those who have been lost, and will be lost, live on. . . .[W]e finally have no choice but to continue deepening our capacity for grief and, through that process, our engagement in the sufferings of the planet. This is the lesson that the beautiful, never-static world perpetually repeats. True grief is never passive, but fulfills itself in the creative imagination. Grief leads us to trace circles where none were visible before, and to understand our ongoing struggles—environmentalist and otherwise—as enactments not of abstract idealism but rather of familial piety.[4]

Environmentalists and concerned citizens are increasingly participating in forms of spiritual counsel and guidance as they explore how to integrate their ecological worldview with the knowledge of the great wisdom traditions. The dialogue between environmentalism and religion continues to expand, reflecting the necessity of adding a moral and spiritual dimension to environmental concerns. A recent anthology by Steven C. Rockefeller and John C. Elder, *Spirit and Nature: Why the Environment Is a Religious Issue*, reports on a conference, an interdenominational discussion of ecology and spiritual questions, held at Middlebury College. The great wisdom traditions provide experience and counsel for issues such as hope and despair, living simply, faith and community, life and death, and many of the questions that pervade environmental work. The National Religious Partnership for the Environment (comprising the U.S. Catholic Conference, the National Council of Churches of Christ, the Consultation on the Environment and Jewish Life, and the Evangelical Environmental Network) in October 1993 launched a 3-year, $4.5 million project to place issues of environmental justice and sustainability at the heart of American religious life. The project involves educational initiatives engaging scientists, educators, and theologians; the development of an environmental curriculum, and even a Green Congregation Hotline, documenting grassroots religious activities. These are remarkable developments, a response to the deep concerns that people feel about environmental issues, and a recognition that ecological well-being is fundamentally a moral and spiritual matter.

The environmentalist faces a profound challenge. In promoting an ecological worldview, he or she, regardless of intention, conveys a tone of moral judgment—advocating ecologically correct choices, behaviors, or policies. Equally important, however, is the public discussion regarding the moral and spiritual implications of global environmental change. Ultimately, behaviors and policies won't be changed unless people are moved at this deep, inner level. The environmental practitioner must learn how to provide such guidance and support. In this way, the environmental profession becomes a healing profession.

When people decide to confront the despair that accompanies ecological decline, they encounter a new phase of ecological identity work. For humans are the agents of global environmental change, and to experience despair, a person must be prepared to assess the cultural circumstances that contribute to global suffering, and consider his or her own culpability. To what extent is an individual responsible for environmental problems? Should blame be attributed to an external other (the culture at large or greedy resource exploiters) or internalized as guilt? Ecological identity work confronts this question, emphasizing how blame and guilt must be transformed into action and responsibility, generating insight and even spiritual transformation. The ability to give voice to these issues is fundamental to environmental education. The next section provides an educational model for approaching this task.

Beyond Blame-Guilt Loops: Taking Responsibility

"When we see photographs and programs about the atrocities of the Nazis, the gas chambers and the camps, we feel afraid. We may say, 'I didn't do it; they did it.' But if we had been there, we may have done the same thing, or we may have been too cowardly to stop it, as was the case for so many. We have to put all these things into our compost pile to fertilize the ground. In Germany today, the young people have a kind of complex that they are somehow responsible for the suffering. It is important that these young people and the generation responsible for the war begin anew, and together create a path of mindfulness so that our children in the next century can avoid repeating the same mistakes. The flower of tolerance to see and appreciate cultural diversity is one flower we can cultivate for the children of the twenty-first century. Another flower is the truth of suffering—there has been so much unnecessary suffering in our century. If we are willing to work together and learn together, we can all benefit from the mistakes of our time, and, seeing with the eyes of compassion and understanding, we can offer the next century a beautiful garden and a clear path.

Thich Nhat Hanh, *Peace is Every Step*, p. 133

The Eco-Confessional

After 20 years of teaching and innumerable discussions about the fate of the earth, I am still frustrated and impatient with a repeating scenario—people jointly bemoaning the overwhelming prospect of irreversible ecological damage. I watch them become increasingly depressed, feeling victimized, paralyzed, either blaming the externalized other or swimming in their own guilt. My frustration merely reflects my own capacity to slip into a similar state of mind. Yet this is a common theme and as an educator it is my role to work a group through these feelings in a constructive way.

I try to do this by facilitating an "eco-confessional. " I ask my students to tell personal stories of ecological irresponsibility. Remarkably, everyone has such a tale. This activity works best with a group that has already built trust and support, based on their common expression of ecological identity. In teaching situations, I only use this activity after I'm convinced that a group will not take themselves *too* seriously and that they are capable of looking at their own actions with mature self-reflection. These stories can lower one's self-esteem if they are used to dig a hole of depression, but they can also relieve burdensome guilt and generate deep compassion and understanding. The purpose is not flagellation, but insight. I find that the collective discussion and interpretation of one's own regrettable actions is a gateway to learning about collective responsibility for the ecological commons. As one example of this approach to ecological identity work I reveal my own eco-confessional (it's *my* book, so it's most appropriate that I embarrass myself!).

For many years, during the last 2 weeks in May, my family has vacationed on a remote island off the coast of Maine. The vacation comes at the end of a long academic semester. It represents a time of meditative retreat, family bonding, and natural history exploration. The boreal coast is a fitting contrast to our customary environment, the deciduous hills of southwest New Hampshire. The open expanse of ocean, the thick spruce woods with their sparse undergrowth, the fog, the play of light on the dappled water, the big sky, the black-and-white flash of eider ducks, the spring warbler migration—these scenes represent a time of opening, opportunities to observe and listen. Our island rental cottage sits perched on a rise overlooking the Gulf of Maine. The house is rustic and comfortable, without electricity or telephone. It has a spectacular view. On one side is the great expanse of ocean. On the other side is a picturesque harbor, dotted with lobster boats, and an occasional fishing trawler.

I spend hours walking the coastline of this small island, reacquainting myself with familiar places, exploring the nooks and crannies of the sea. I have many favorite spots, where I sit for long stretches of time, emptying my mind, observing nature. I have one particular "sitting" spot, just a few feet from the cottage. It's a subtle hole, shaped so that it fits the contour of my body perfectly. It overlooks the ocean and the harbor. This is a timeless place, ideal for contemplation, intrinsic to my ecological identity.

Many years ago, when my daughter was an infant, we had to decide whether to bring disposable diapers with us to the cottage. On the island, we have to wash all of our clothes by hand. Since there is no electricity, a limited supply of propane, and a possibility of consecutive days of inclement weather, washing clothes is inconvenient. Against our better judgment, we thought life would be easier if we used the disposables. We brought a good supply of these diapers (an entire duffel bag) with us from the mainland. At that time, there was no recycling on the island. Garbage was either burned or tossed into the ocean. Knowing this, we made every effort to be mindful of our waste, disposing of it in the most ecologically responsible way possible. But we had one major problem: the growing accumulation of used disposable diapers. The ecologically responsible path was clear: separate the plastic, throw the waste into the ocean, and bring the plastic back to the mainland. But this was very inconvenient and extremely messy. On a small island, without formal garbage removal, the disposable aspect of the diaper became a significant concern.

After some discussion and several inquiries about how the local population deals with this issue, we chose the easy and lazy method. I would put the diapers into large, plastic garbage bags and throw them into the ocean. I was hoping for a windy day, so the bag would quickly float out to sea and then become submerged so it couldn't be seen. But the ocean was unusually calm the day of our departure. Assuming that I had collected the last of the disposable diapers, I threw two large garbage bags, completely filled with those wretched things, into the calm, pristine, greenish-blue waters of the Gulf of Maine.

Several minutes later, seagulls pecked open the bags, scavenging my daughter's fecal matter. Before long, dozens of diapers were floating across this beautiful harbor, forming a white path of plastic waste. They were like white buoys on the water, meandering in the calm harbor, slowly spreading out, marking a trail of neglect.

"Gosh, I hope no one else sees all of this," I said to my wife. "We'll probably be banned from the island forever. Hopefully, they'll wash out of sight, so no one will ever know where these things came from."

But as we had the only infant on the island it was pretty obvious who was responsible for the mess. I'm not sure anyone who observed this spectacle cared very much. People routinely threw all sorts of garbage into the ocean. Nevertheless, I was profoundly embarrassed, suffering enormous shame, feeling like a first-class jerk, and a moral hypocrite.

Imagine a group of dedicated environmental studies students telling similar stories. The collective discussion becomes silly and profound. The diaper story is merely a disturbing symbolic example of ecological irresponsibility. People recall wasteful energy use, redundant consumer purchases, joyrides in their cars, and so on, laughing at themselves but also seriously considering the meaning of their actions and the lessons they might serve. The stories are a reminder that everyone contributes to pollution. Often people dump what they no longer need and don't consider the consequences. Out of sight, out of mind. Good recycling notwithstanding, many environmentalists, like most citizens, are chronic dumpers. We look closely for the signs of the polluter. We have found the enemy and it is us.

Surely there is also a cultural and institutional context for dumping. There are people who sanction dumping on an enormous scale, large enough to wipe out entire habitats. For example, the Russians continue to dump hazardous radioactive waste into the ocean, reassuring the world that the impact is minimal. Oil spills typify negligent dumping. The oil companies, the regulators, and the public all know that the risk is there, but the supertankers still sail, and accidents continue to happen. Throwing disposable diapers into the Gulf of Maine is not equivalent to dumping dioxins in a river, but there is something that connects both actions. We cannot just blame other people for the environmental crisis. All citizens are participants in a nonsustainable system.

Blame and Guilt

In the last 30 years, much has been written about the dark side of industrial civilization. There are stinging critiques of affluence, linking the imperial leviathan of states, corporations, technology, and belief systems in the service of progress and economic growth, if not personal greed and power. Although this is well-trodden territory, and I will not go over it again, it should be understood that almost every environmental studies student at some point learns that America uses a share of energy and resources that is unconscionably out of proportion to its percentage of the world's population; that transnational businesses, many of which are based in the United States, are responsible for an extraordinary amount of environmental pollution; that economic growth, for all of its material benefits, has wreaked ecological havoc; that various development schemes have disrupted indigenous cultures, and so on.[5] At some point, many people recognize that they are reflections of a culture that has perpetrated environmental deterioration. With only a modicum of self-reflection a person must ask: To what extent am I responsible for all of this? Is this my legacy as well?

There are distinct stages that people experience as they internalize the magnitude of environmental deterioration. First, they blame the perceived perpetrator—greedy industrialists, imperial nation-states, misguided worldviews—the externalized other. It is their fault, and we are all victims of their actions. But upon closer examination, they begin to realize their own culpability. If I am a member of the culture, and I contribute (unknowingly perhaps) to its actions, then I am also the perpetrator. This leads to feelings of guilt and shame. A common, unhealthy dynamic is the perpetual victimization of the blame-guilt loop, moving back and forth from the "perpetrator without" to the "perpetrator within." What allows people to move through these stages so they can take responsibility for their actions and move forward to change themselves and society? Herbert Fingarette, in *The Self in Transformation,* interprets the psychodynamic and spiritual implications of blame and guilt, suggesting how these feelings can be transformed into responsibility and action. His insights are of great interest to the environmental practitioner.

One's first impulse, when hearing more bad environmental news (an oil spill, a nuclear accident, etc.) is to determine who is responsible for the injustice. Fingarette describes the conscious features of the blame experience:

> (1) the emotional, quasi-pleasurable brooding upon the other's wrongdoing, (2) the increase in the sense of self-esteem and self-righteousness, (3) the moralistic attack upon the wrongdoer, and (4) the resulting sense of catharsis. The inherently involuntary, spontaneous character of the impulse to blame is accounted for by the fact that the process depends upon the existence and mobilization of inner conflict which remains unconscious.[6]

When the *Exxon Valdez* ran aground on Bligh Reef in Alaska's Prince William Sound, spilling 11 million gallons of oil, many people were shocked at the devastation and negligence. Their first impulse was to find a scapegoat and blame the responsible party. The news media went to work, trying to assess who was at fault, trying to find the dirtiest laundry. They discovered that the captain of the vessel may have been drunk. Aha! It was his fault. Then they looked at the Exxon Corporation to determine its culpability. How do they manage these tankers? Why was the ship sailing in such rough waters? How responsible was Exxon? Certainly Exxon took a hefty portion of the blame. By blaming Exxon, people initially felt better about themselves. Environmentalists self-righteously proclaimed that this was an accident waiting to happen. Public attention was focused on the perpetrator, allowing people to discharge their anger, perhaps temporarily relieving their anxiety, and avoiding the inner conflict regarding their own culpability.

Certainly a portion of blame is necessary. It is important to locate the perpetrators and hold them accountable for their irresponsible actions. Blame serves as a transitional state for the psyche. But it is necessary to move through blame and try to understand the source of the conflict and tension. Blame allows people to deflect anxiety and temporarily relieve themselves of stress, but it is never a final solution. Blame may be justified, but it should not be an end in itself.

To what extent are all American citizens, including environmentalists, responsible for the *Exxon Valdez* ? Exxon should be held accountable for their actions, but that doesn't relieve "distant observers" of their responsibility or their connection to the disaster. For example, after the oil spill, some analysts, many of whom were environmentalists, called attention to the type of economy and the cultural orientation that allows such dangerous ships to be built in the first place. Perhaps the whole culture is at fault and all people who drive cars to some extent share the responsibility. What are the historical, economic, and political conditions that contribute to the *Exxon Valdez* ?

In my teaching, I encourage my students to ask these types of questions. For example, as environmentalists become more aware of the consequences of global environmental change, they develop interest and concern regarding the plight of indigenous cultures and the fate of the world's tropical rain forests. I juxtapose two historically distinct but conceptually linked examples of how a landscape is ecologically ıgs cannot be avoided and must be confronted. But how can these ıgs become part of a healing process, seen in the full context of ɔgical identity work, used as an educational means for insight and ɔnal growth?

ı Guilt to Responsibility

arette describes guilt as a transitional state, explaining that the nce of both therapeutic and moral progress is the ability to ɩowledge feelings of guilt and to accept responsibility for doing ething about the issue one feels guilty about.[8] Otherwise one suf- both the moral and psychological consequences of living in the ɩe-guilt loop.

ɔr example, on a simple, interpersonal level, two friends may have stressing argument. Afterward they both reflect on their behavior realize that they used inappropriate language causing hurt and ppointment. They both feel guilty about what they said. If they are illing to take responsibility for their words, they can easily talk ıselves into believing that it was something the other person said that caused the disagreement. Then it is the other's fault, and he ıe is blamed; it was something he or she did. The blame-guilt loop finite until the cycle is broken. If the friends take responsibility for actions, recognize how they each contributed to the misunder- ding, then they can reenter the conversation, without blame and :, and achieve a greater understanding of their situation.

onsider the same dynamic on a broader cultural level by returning ɩe diaper story. I can abnegate my responsibility for tossing diapers the Gulf of Maine by thinking that there was no recycling struc- and I was just doing what everyone else on the island does. I alle- ɛ my guilt by blaming the social structure at large. Or I can trans- ı my guilt by taking responsibility for my actions and trying to struct a better alternative. This question becomes more difficult abstract when we consider situations that seem further removed ı our daily affairs. After all, I can take responsibility for dumping ɩers in the Gulf of Maine or for cleaning up my own life, but is it fault that the Europeans brought smallpox to North America? I ı't create the vast global supply of plutonium or build the super- :ers. Why is it my responsibility to do something about those prob- s? Isn't that an unjustifiable burden?

and politically transformed. We study the Penan, a
that has experienced unspeakable dislocation and
expense of corporate and statist tropical logging. W
about this, they are shocked and horrified at the rapi
ical deterioration and cultural disintegration.

Then we consider the transformation of the Ne
scape, observing the land under our feet, the pla
inhabit. Although the historical circumstances are di
formational processes are similar. In New England,
tion of nature resulted in the marginalization and
indigenous culture, the destruction of many indige
ecological reconstruction of the landscape, the expl
resources, the spread of harmful pathogens, and fina
ization of the region. Carolyn Merchant, in *Ecologic*
resents this as the repeating cycle of global envir
Students respond to this material emotionally, wi
side of economic growth and industrialization, loo
lives in contrast, wondering how they are person
these impersonal forces.[7]

When my students confront the historical circum
ing to ecological decline, they often internalize these
ring the blame to themselves, experiencing a stro
The issue is personalized: suddenly it is their cult
and lifestyle that are at fault. This mess is their resp
suffer shame and embarrassment. How could my a
in such a destructive manner? How could they ha
ful? It is wrong of me to blame others. I must also bl

People are easily overwhelmed by the magnitud
mental change, plummeting into an existential d
whether there is any way out, if there is anything
rectify the situation, to alleviate the anxiety, to mov
and blame. This is where the blame-guilt loop m
victimized and exploited by a situation that is out o
was unexpected, or for which someone else was i
This casts a disquieting shadow, becomes a place
ing, in which people shift from blame to guilt to
take action, feeling self-pity, and plagued by doub
moved to action, they are immobilized by their gui

It is understandable that people blame others o
al guilt when they acknowledge their own culpabi

But as Fingarette asks, "What is the (moral-therapeutic) *solution* to the present human predicament, granted that what happens now is a consequence of what happened when we could not control what happened?"[9] By looking deeply at the present, we inevitably review the past. Just as we often blame our parents for the mistakes they made, we might blame the English settlers for destroying New England's indigenous populations. As our biological parents are responsible, in part, for how we grew up in the world, so our ancestors are responsible for the historical and ideological circumstances of our contemporary situation. We may not have control over the mistakes that they made, but we can avoid those mistakes in the future if we take responsibility for our future actions.

Many people are in denial regarding their social and environmental responsibility. They abnegate their responsibility by claiming they have no control over the events that contributed to their plight. The path to maturity and insight, both as individuals and as members of a community, is learning how to take responsibility for events over which we seemingly had no control. In this case, ecological identity work broadens the circles of identification to include the historical legacy of our ancestors, and the intangible interconnectedness of personal and collective responsibility.

Guilt is retrospective, enabling us to consider the consequences of our past actions. Responsibility is prospective, inspiring positive action to construct the world as we would like it to be. It is through our present action that we incorporate the past and acknowledge the future. Guilt and responsibility become the dynamics of spiritual transformation, allowing us to merge the suffering of the past with the liberation of the future. This is a fragile balance, precariously holding what happened before with what will happen tomorrow, juggled in the hands of today's life choices.

In this context, we can interpret our "eco-confessional" stories, uncover what had been denied, relieve the burden of subsequent guilt, acknowledge responsibility, avoid thoughtless choices in the future, work harder to understand the conditions and structures that lead to ill-conceived action, discuss these situations with other people, and develop collective responsibility for our interconnected transgressions.

The ability to take responsibility for our actions is a prerequisite for ecologically responsible citizenship. It is also a therapeutic action, contributing to a process of personal and political healing. We cannot change the past, but we can strive to improve the future. We cannot rewind the tape on 400 years of European expansion, but we can understand how historical circumstances inform contemporary issues. By studying the past, by understanding the repeating cycles of ecological transformation, we can work with integrity and compassion to deal with these cycles as they emerge in contemporary experience. For Fingarette this requires the perspective of humility:

> Honest humility reveals that to accept responsibility, considering what we start with is a heavy burden. To say, as criticism, that this is not "fair" or "just" is to suppose that the world is fair and just. This is precisely what the world is not. It has no design leading to some inevitable, built-in moral future. It is we human beings who can reach humanity only by accepting the challenge to make the world just.[10]

By taking responsibility, we respond according to our best moral judgment and our deepest compassion. There is a "healing" component to responsibility, the therapeutic and moral imperative to construct a viable future, to contribute to the ecological commons, to act according to our ecological identity. And with such constructive action, we hold the fate of the world in front of us, taking responsibility for the future as we cultivate insight about our lives.

Ecological Identity and Reflective Practice

Stress and the Environmentalist

During the spring of 1991, the Natural Resources Defense Council, the Buddhist Peace Fellowship, and the Nathan Cummings Foundation co-sponsored a meditation retreat for environmentalists. Thich Nhat Hanh, the Vietnamese Buddhist monk and peace activist, was the teacher. The retreat combined meditation, silence, dharma talks (spiritual and philosophical lectures), presentations, and small group networking. The point was to allow environmentalists to look more deeply into their work and to provide a reflective perspective on environmental action.[11]

The idea of "taking care of the environmentalist" was an effective integrating theme. The retreat involved much silence and sitting, but there were also opportunities for people to talk. Since discussion time was limited, people poured out their ideas, realizing how many things they wanted to talk about, and how much they had in common. Many people cited the various forms of burnout they experience: activists tired out by continuous political confrontations; administrators made weary by the constant pressure to raise money; educators overwhelmed by the profound attitudinal changes that are prerequisites for environmental reform; and concerned citizens feeling despair at the magnitude of environmental loss.

The overriding theme of the retreat was that environmentalists, in their professional zeal and personal commitment, were overworking themselves, failing to adequately nourish their ecological identity, abnegating their responsibility to their own health and well-being. Thich Nhat Hanh brought this perspective to his dharma talks, couching it in the language of "healing."

> They [environmentalists] must learn to take care of themselves. Activists need strength, especially spiritual strength. You can only share success or happiness when you have these things in yourself. So in order to take care of the environment, you have to take care of the environmentalist. Activists who want to protect the Earth have to learn to protect themselves. This retreat is an opportunity for them to have the time to look deeply. Those of us who are wounded in the process of serving need to learn the art of healing ourselves—and to learn it from each other.[12]

But what does the idea of "healing ourselves" really mean? Who has the time to do this, or to make it a professional and personal priority? The retreat participants would all return to their "busy" organizations, confronting the same daunting tasks: workloads, meetings, personnel issues, budgetary restraints, and so on. And they would return to the reality of the ecological crisis. Their lives would still be driven by the same stressful demands, the urgency to achieve ambitious agendas, the feeling of having to solve one crisis after another.

Yet the clear message of this retreat was for environmentalists to reconsider how they approached their work, to engage in a more reflective professional practice, buoyed by a commitment to personal introspection. Environmentalists have to care for themselves, not only for the obvious reasons of self-preservation and personal growth but because they serve as role models for sustainable living. Many environmentalists live a life of contradiction. On the one hand, they promote an environmental way of life, which involves appreciating the joys and wonders of living sustainably on the earth. On the other hand, they endure stressful lifestyles, pushing their organizations and themselves to unreasonable limits.

The environmental practitioner has to deal with many levels of stress. There is the personal stress faced by many modern professionals: the ambition to achieve and accomplish, the need to be successful, the ability to juggle multiple roles, balancing family and work, stretching the day so more can be accomplished, or finding time for contemplation and recreation. There is the organizational stress of living with tight budgets, understaffed projects, and pressing deadlines. There is the moral stress of having to live up to the goals and ideals of an environmental way of life, becoming a role model for others, demonstrating practical alternatives. Finally there is the awareness of global environmental stress, knowing about all the different ways that the planet is in distress.[13]

Consider a typical day in the life of an environmental administrator. The morning is spent analyzing the scientific data necessary to evaluate a piece of state legislation. The afternoon is devoted to a seemingly intractable administrative problem—the latest budget cuts facing the agency. In the evening, it's off to a public forum where engineers, politicians, and local residents will discuss a controversial wastewater treatment plant. And that's just Monday's agenda. Tuesday will bring another set of challenges: staff meetings, public appearances, fundraising, advocacy, and so on.

Or consider the challenges facing an environmental educator who is working on the front lines at a successful nature center. The morning is spent in a local public school, working with school children, presenting a program about hawk migrations. The afternoon is devoted to developing a hazardous waste curriculum, an approach that incorporates science and citizenship, that teaches students the complexity of waste issues while emphasizing the importance of citizen participation. In the evening, there is a public presentation on the latest plans to bring a shopping mall to town. Our educator is a member of the local citizens group, but has to be wary of citing her nature center role, so as not to jeopardize the group's reputation for providing a balanced approach to controversial issues.

Finally, consider a day in the life of an environmental technician, who is on the front lines cleaning up a toxic waste dump. He is not sure that this type of work is his life ambition, but the pay is good, and this is what he has been trained to do. On Sunday night, in the airport hotel, he lies awake contemplating the next day's work. On Monday morning, he is driven to the pollution site. He puts on his protective coat, gloves, and mask. He spends the day cleaning up this extensive field of sludge and grime, mining the pollution, performing the ultimate custodial job, a member of earth's industrial janitorial service. In the evening he returns to his hotel; he watches television and falls asleep.

Many environmental professionals are drawn to the profession for moral and spiritual reasons, inspired by their experiences in nature and their awareness of the environmental crisis. Yet often they ascend to leadership positions for organizations that are in perennial crisis, or they wind up with indoor jobs, and their ecological identity feels like a dream, or something they pursue in their spare time, when they're not busy with the stressful circumstances of professional life. They find most of their time is spent dealing with the innumerable management issues that their organizations face. Not only are they involved in a type of work that is removed from their original professional goals but they are not "trained" as managers. They work for small or moderate-sized organizations where one person plays multiple roles and wears many hats.

In order to "take care of the environmentalist" they have to learn how to take care of their organization, finding ways for their organization to cope in turbulent working situations so that they can accomplish the environmental mission of the organization. A recent study, sponsored by the Conservation Fund, emphasized the severity of this management problem.

> In their day-to-day operations, many conservation-environmental groups are so busy struggling with finances, overbearing work loads, and staff teetering on the brink of burnout that they can hardly find the time or the will to look for long-term solutions to their problems. Two crucial solutions are especially underused: planning and training. Caught up in crisis management, organizations defer long-range planning. . . . Most conservation leaders learn organization management in the school of hard knocks. The majority of conservation groups are perpetual management experiments.[14]

Hence many people in these organizations fall into a trap in which they become so task-oriented, so driven by the enormity of the issues at hand, that they don't take the time to think about the organizational process through which they make decisions, or they don't talk about what it means to be an environmentalist, or most ironic, they don't have the time to commune with nature, the very motivation that inspired their participation in the profession. This type of organizational stress is faced by many environmental professionals.

Reflective Environmental Practice

In effect, Thich Nhat Hanh's message at the environmentalist retreat was that practitioners cannot implement a sustainable society unless they learn how to cultivate sustainable psyches. These are parallel "healing" processes, the ability to restore an ecosystem and the personal awareness to restore one's psyche. An overdeveloped, polluted, disturbed ecosystem is no different from an exploited, burned-out psyche. Both require the full attention of the reflective environmentalist.

Educators refer to "reflective practice" as an approach to professional training that enables the practitioner to use self-awareness, critical evaluation, and interpretive observation as introspective learning tools. The goal of reflective practice is to understand the consequences of professional action, to assess how one is perceived as a practitioner, to use professional activities as an educational laboratory to learn about the issues of the profession, and to connect those activities to one's value system and personal growth. Hence "reflective practitioners" always consider the broad context of their work, placing particular emphasis on the learning process of professional activity. They construct and integrate a professional and personal vision, stepping out of their work to consider whether their actions conform to that vision.[15]

Many environmental practitioners become so involved in their work that they lose this reflective perspective. Their professional identity becomes paramount as they identify with their job descriptions, careers, professional roles, and organizational tasks. As an educator in an environmental studies program that trains practitioners, my objective is to integrate ecological identity with professional identity. Why do many practitioners get so involved in their work that they no longer have the time to nourish their ecological identity?

In part, this occurs because environmental organizations, like many social service organizations, are strapped for resources. Staff and organizational retreats are not a priority when dozens of other urgent tasks seem more important: raising next year's funds, the business of the current legislative session, taking care of the visiting school group, and so on. Yet in the same Conservation Fund study cited earlier, when executive directors were asked about the factors that could make their work more rewarding and effective, three of the top five answers were more time for myself, more opportunities for personal growth, and more opportunities for professional growth. To be sure, the same sentiment exists among the staff.[16]

When I work with a group of environmental practitioners, and with one voice they protest the stress of work responsibility, I involve them in a brief, but telling reflective exercise. I ask them to create a series of pie charts. The first chart describes how much time they spend doing various activities, i.e., sleeping, eating, working, playing, etc. The second chart divides their work time into proportional segments. The third chart is an idealized portrait of how they would most like to spend their time. The most common recurring pattern is that the time spent involved in outdoor recreation, the contemplation of nature, or work on personal growth is relegated to a narrow sliver, although it represents a hearty chunk on the idealized chart. I listen to the full gamut of excuses and rationalizations: "As I gain more work experience, I am given more administrative responsibility and I have less time in the field and more behind a desk." "As I develop my expertise, more work flows in my direction. I can't turn it down because it's such important environmental work." "My organization has grown over the last few years and the demands on my time have grown accordingly."

I find that staff and leadership are hungry to talk about the relationship between their ecological identity and professional practice, but they are not sure how to. Many organizations do an excellent job as role models in areas like political action, public programs, and environmental consulting, but when it comes to confronting issues of personal and organizational stress, or considering the psychological and spiritual impact of environmental work, they are pretty much in the dark.

Many executive directors have told me that they would love to have more time to participate in ecological identity work, that is, to experience time outdoors with their staff, to talk together about what it means to engage in environmental practice, or to consider ways of dealing with stress and improving the work environment. But they find that there is just not enough time in the day to accomplish this. They are sure that such discussions are important, but wonder whether such deeply personal questions may open a complex web of psychological issues, requiring delicate and experienced facilitation.

Environmentalists constantly find themselves in situations where their values are being challenged, where they have to confront issues of loss and despair, where they have to make personal sacrifices for the good of their organization or community, or where they confront other people with difficult decisions. During these times they have to dig most deeply, find their reservoirs of strength, find networks of support and encouragement, find ways to bring people together, summon clarity and vision. The best technical expertise, although helpful, will not guide people through these challenging circumstances. And as people become more skilled in their work and more active in a community or organization, as they attain more leadership and responsibility, situations such as these increasingly occur.

When people face challenging decisions, do they have the reflective capacity to look deeply at the ramifications of their choices? This is an educational imperative, to provide people with the orientation and experience to confront moments of risk, ambiguity, moral conflicts and controversy. When practitioners are willing to confront these matters, both publicly and privately, then they take seriously their role as healing practitioners. This is the kind of educational leadership that the environmental profession requires. These are the issues that many practitioners face on a daily basis.

Depending on the size and mission of the organization, there are steps leadership can take to allow staff to confront some of these issues. These include periodic retreats affording the quiet and privacy in which people can discuss the difficulties and challenges of being an environmentalist; opportunities for discussing some of the moral dilemmas that occur when staff participate in controversial public issues; discussions of the various value conflicts that people confront on the job; and opportunities for staff to take short natural history excursions together. These are means to incorporate reflective practice into the daily life of the environmental organization.

The challenge for the reflective environmentalist is to implement such approaches in workplace settings. Perhaps one path for doing so is to build such approaches into organizational mission statements. Environmental organizations have a major role to play in bringing these reflective issues to the forefront of their public work. If an environmental organization is to grow and learn, it must facilitate reflective practice, among both its staff and its clientele. This inspires invigoration and change, and allows the workplace to become a community, integrating professional practice and ecological identity.

An environmental organization cannot be successful unless there are shared goals and objectives, a willingness to work together, and an understanding of the personal and professional needs of staff and leadership. It is a place where people learn about their profession, the important public issues that surround their work, the way other people perceive their ideas, where ideas are shared and communicated or where various group decisions get made. Environmental organizations have an educational obligation to their staff and clientele to recognize the moral and spiritual magnitude of their mission, to incorporate reflective discourse as part of the job description.

Taking Care of the Environmentalist

People often say "take care" as a way of saying good-bye. Are these words just a manner of speech or are they a serious statement about the meaning of life? In one sense, "take care" is a polite formality, bringing closure to a social interaction. But these words imply a powerful and profound message.

To "take care" of somebody or something implies nourishment, support, stewardship, and healing—acting with deliberation, mindfulness, and understanding. The word *care* has several etymological roots. It means "to grieve and lament," but it also means "to have affection or liking for." This corresponds closely with the moral and spiritual message of environmentalism, simultaneously grieving and bonding, accepting loss and love. If I say that I care about my family, community, land, or planet, I declare a sense of belonging and identification, acknowledging the importance of the things that I care about, taking responsibility for their well-being.

What does it mean to "take care of the environmentalist"? My interpretation is that people take responsibility for their decisions, are mindful of their own well-being and the health of their community (from landscape to ecosystem to planet), and understand the broad context of their actions. It means being attentive to one's moral concerns and spiritual needs, implying thoughtfulness, patience, deliberation, and concern. On a practical level, "taking care of the environmentalist" represents a commitment to remove the layers of personal and global stress, exploring one's inner life, contemplating the wild, providing time for work *and* play—all aspects of exploring ecological identity.

Above all, taking care represents the pursuit of happiness, which is intrinsic to ecological identity and healing. In a small pamphlet entitled "Compassion and the Individual," Tenzin Gyatso, the Fourteenth

Dalai Lama asks what is the purpose of life, the one great question which underlies all human experience.

> I believe that the purpose of life is to be happy. From the moment of birth every human being wants happiness and does not want suffering. Neither social conditioning nor education nor ideology affect this. From the very core of our being, we simply desire contentment. I don't know whether the universe, with its countless galaxies, stars and planets, has a deeper meaning or not, but at the very least, it is clear that we humans who live on this earth face the task of making a happy life for ourselves. Therefore, it is important to discover what will bring about the greatest degree of happiness.[17]

The Dalai Lama suggests that inner peace is crucial to happiness and that "the greatest degree of inner tranquillity comes from the development of love and compassion." So taking care of the environmentalist means to cultivate love and compassion in order to attain inner tranquillity. The happiness of inner peace is difficult to attain. It takes guidance and support. Most environmentalists find this happiness when they participate in nature, when they feel integrated with their environment, when they are filled with gratitude and wonder. This means learning how to love the earth, undeniably the greatest challenge for all environmentalists. If ecological identity enables people to identify with the earth, then to love the earth is to love oneself. This is how to take care of the environmentalist, making this orientation intrinsic to daily experience, the goal of personal and collective healing, the task of every environmental practitioner.

6 Educating for Ecological Identity

The most exhilarating moments in teaching occur when an inspiring learning community emerges from the structure of the course. There are situations when the life experiences of the students take over, and they relate how the circumstances of their lives, their observations, interpretations, and reflections come to bear on the material. The class takes on a life of its own. As a "teacher" I can move to the background, listening to the insights of the class, letting the material speak for itself, watching a collaborative voice, an evolving text, a creative spirit permeate the educational process.

This situation is most likely to occur when the classroom becomes a cohesive, dynamic community. A class is a temporary community in which people learn together by looking deeply into their lives, an opportunity to deal with important issues, a place to discuss the meaning of life, a place to practice thinking about the meaning of environmentalism. It is a great privilege to come together to think about nature, to participate in ecological identity work, and to do so with other people who have similar goals and aspirations. As the weeks of a class roll by, the community strengthens, the collaborations deepen, people begin to realize how much they have in common. The vitality of the learning community is as important as the pursuit of knowledge.

I think of myself as an educational designer and when I organize a course, I contemplate the experiences I want to have with my students as we become engaged in the material. Out of these experiences the content will emerge. I am always searching for designs that will enable my students to understand the depth of their personal experience. For when this occurs, they are integrated learners. They inspire those around them. They become, if you will, *sacred learners*. And it is through this process that I too am opened up to learn more about the material and more about myself. This is the educational orientation that guides my teaching experience. It provided me with the material and guidance to write this book.

Ecological identity is above all an educational process, an approach to learning that integrates citizenship, professional practice, and personal growth. I have provided numerous examples of educational designs, activities, and applications, intended to develop vital and vibrant learning communities, an educational commons in which people thrive on the quality of their learning experience. When people participate in ecological identity work, they see the expansive realm of reflective environmental practice. This is a collective effort—introspection for the purposes of ecological citizenship, personal awareness to promote common responsibility, mindfulness to expand understanding of human/nature interactions. When a group of people work together to explore these relationships their ideas emerge simultaneously, revealing a matrix of interpretations, a spectrum of insight, an illuminated learning community.

This chapter places ecological identity in a broad educational context, describing how it serves as a framework for the teaching of environmental studies, a basis for establishing profound learning communities, and an approach to lifelong learning. The first section suggests that ecological identity is the epistemological glue for reflective environmental practice, integrating the formal education of professional training with the learning experiences of everyday life. Environmental education should be concerned not only with what people know but how they learn. Using David Orr's interpretation of ecological literacy, I emphasize how the *quality* of knowledge is critical to the educational process. The second section provides an outline of the teaching techniques that are conducive to ecological identity work. It serves as a handbook of educational design and is intended for the professional educator. How does the educator create learning structures that promote collective vision and personal growth? I suggest specific design criteria that attend specifically to the learning process, criteria of direct relevance to adult learners, but useful in a variety of settings. The third section presents the sense-of-place map activity, an educational rite of passage for ecological identity work, an approach to learning that synthesizes and integrates many of the themes of this book.

Ecological Identity as a Framework for Environmental Studies

For the last 20 years, I have spent the majority of my professional time designing curricula, planning programs, advising students, visiting practicum sites, working closely with environmentalists in a variety of professional situations, in an effort to provide the most meaningful and applicable graduate school experience for environmental practitioners. This process has been dynamic and improvisational as I have watched the profession change and evolve with the increasing complexity and pervasiveness of environmental problems. I have had hundreds of conversations regarding what environmental practitioners should know and how their education should be organized, both in terms of their personal needs and the needs of the profession. The challenge has always been to provide a program that has a unified educational vision, flexible enough to meet the needs of different types of learners with a variety of professional goals.

What do the hazardous waste policymaker, the nature center director, and the elementary school environmental educator have in common? What educational processes link them together? It has become increasingly obvious that common educational ground is found not so much in what people know but in how they learn. Ironically, as the environmental profession becomes more specialized and the knowledge requirements are differentiated, it is the *learning process* itself that can serve to integrate the profession.

As an approach to environmental studies, ecological identity work contributes to a vision of reflective environmental practice, a holistic interpretation, grounded in real-world problems, applied to the challenging environmental issues that demand the practitioner's attention. Environmental studies students at any level (K–12, college, graduate) or in any setting (museum, nature center, classroom), whether they are training as practitioners, or are just concerned citizens, are interested in applying their knowledge in order to protect nature and promote environmental quality, but that work is hollow unless it also corresponds to their deepest values about nature. Ecological identity work is not only intended for personal growth and awareness, it is a framework for ecological citizenship, and the educational basis for reflective environmental practice.

This section stresses the importance of ecological identity as both a conceptual foundation of formal education and the experiential curriculum of everyday life, explaining the function of reflective learning generally, placing it within the broader educational goals of environmental studies, linking it to ecological literacy, and showing its practical application.

Content, Process, and Reflection

Environmental education, as it applies to all aspects of environmental studies, must strive to integrate three interconnected domains of knowledge: content, process, and reflection. *Content* is the information that flows through a system, the relevant phenomena of a system or object of study, the extrapolation and observation of relevant data. *Process* refers to the ways that people share and use information, the relational context in which learning occurs, the way information is represented. *Reflection* is the personal or collective interpretation and contemplation of information, its psychospiritual implications, its deep meaning. These are dynamic, fluid categories, composing an integrated approach to learning.

In a field as wide-ranging as environmental studies, there will always be curricular debates about the most important content, what is often referred to as the "knowledge base" of environmental studies. As the range and depth of environmental information becomes increasingly specialized and complex, discussions about the most appropriate knowledge base become even more controversial.

One cannot know everything, so what is that everyone should know? Depending on the specific orientation of the learner (career plans, interests, proclivities, etc.) the right formula will be idiosyncratic. Understandably, teachers and students alike will have strong opinions about what environmental studies students should know. Suffice it to say that any content mix should be reasonably interdisciplinary and include the necessary ingredients for ecological thinking, which involves an understanding of ecological principles, investigation of the metaphorical implications of ecological relationships, and the application of ecological principles to the human sciences and humanities. The theory and practice of field ecology and environmental science are appropriate building blocks, providing an analytical understanding of ecological systems. Most critical is the learner's ability to observe complex systems and learn how to extrapolate relevant information and patterns, distinguishing the substantive concerns. This can be achieved at any educational level, as long as the corresponding subject matter is developmentally appropriate.

Equally important is the process knowledge that enables people to convey information, to engage in collaborative problem solving, to communicate what they know, and to resolve differences. How do people exchange information? Skills such as community building, communicative competence, interpersonal collaboration, cooperative learning, and democratic participation enable people to understand themselves in relationship to others, connecting them to the workplace, the neighborhood, the school, and the community. Often process skills are considered as a form of technical expertise (marketing, public relations, graphic design, presentation techniques, computer skills, etc.), but they also involve the ability to understand group process, organizational life, and social interactions. Environmental practitioners and ecologically responsible citizens must know how to communicate and collaborate with other people, otherwise their content knowledge will be of no use to the community.

A field ecologist may have completed a thorough, sophisticated study of a nearby wetland, but if she or he can't adequately present the information to the relevant community, it remains locked away. A team of consultants from different organizations and universities may be contracted by a state agency to develop ecological guidelines for the regulation of toxic wastes, but if they can't work together, the study will be much less effective. Technical expertise without process knowledge is not particularly useful. On the other hand, there are people with extraordinary relational skills, but they have no idea what they are talking about. The person who excels in both domains is most likely to exercise environmental leadership and be a successful practitioner.

Reflection represents the psychospiritual implications of content and process, how what we know changes the way we think, how content knowledge changes how we perceive the world and think about the meaning of life. Reflection involves mindfulness, introspection, and deliberation—thinking carefully about the personal meaning of knowledge, considering the wider ramifications of personal and collective action, and using information and relationship to attend to the moment, the direct experience of the here-and-now, the direct experience of nature. This reflective capacity is the core of ecological identity work—the integrating capacity to make knowledge whole.

This reflective, ethical approach is often relegated to the academic work of theology schools, religious institutions, and environmental philosophy programs, or seen as the popular province of New Age workshops and fringe environmentalism. Incorporating this approach with environmental studies is an important challenge for our institutions of learning. Ecological identity work has its place in the business school, the engineering curriculum, or any place where environmental studies is taught.

Environmental issues represent complex values dilemmas, challenging people to make difficult decisions. In these cases, what does a person rely on—the equations of costs and benefits, the calculus of political expedience, the psychospiritual reflection that brings deep inner strength and wisdom to the decision-making process, or a judicious combination of all three approaches? This is the purpose of ecological identity work—the ability to connect substantive environmental issues to the challenges of work and life, through the dynamics of reflective learning.

Environmental practitioners and concerned citizens have extraordinary demands on their time and their psyches. The necessary attributes for being a "successful" professional or citizen are formidable and complex. People require creativity, imagination, and analysis to develop new frames of reference, salient metaphors, and innovative problem-solving techniques to meet the challenges of environmental reform and to promote an ecological worldview. They must understand the diversity and complexity of organizational life, political behavior, and collective decision making. Most important, they need the patience, skill, foresight, and ethical thoughtfulness to deal with the ambiguity, uncertainty, and complexity of environmental issues. This is the goal of reflective environmental practice—the ability to integrate intellectual, emotional, and spiritual faculties in a variety of roles and contexts, to cultivate what David Orr calls "ecological literacy."

Ecological Identity and Ecological Literacy

In his important and thoughtful book, *Ecological Literacy*, David Orr emphasizes the importance of "rethinking both the substance and the process of education at all levels." He describes six foundations of education, relevant not just for the teaching of environmental studies, but for the larger human purpose of learning how to live sustainably:

- All education is environmental education.
- Environmental issues are complex and cannot be understood through a single discipline or department.

• For inhabitants, education occurs in part as a dialogue with a place and has the characteristics of a good conversation.

• The way education occurs is as important as its content.

• Experience in the natural world is both an essential part of understanding the environment, and conducive to good thinking.

• Education relevant to the challenge of building a sustainable society will enhance the learner's competence with natural systems.[1]

What is so impressive about Orr's work, and what makes it relevant for ecological identity, is the emphasis he places on the *quality* of knowledge, and his ability to integrate the three domains of knowledge described above—content, process, and reflection. He provides solid suggestions with regard to the content of ecological literacy and he places as much emphasis on the learning milieu, the places where people learn about nature, the importance of direct experience, the clarity of a reflective orientation. For example, Orr shows how a good conversation is an appropriate metaphor for the teaching of environmental studies.

> Good conversation is unhurried. It has its own rhythm and pace. Dialogue with nature cannot be rushed. It will be governed by cycles of day and night, the seasons, the pace of procreation, and by the larger rhythm of evolutionary and geologic time. . . . The form and structure of any conversation with the natural world is that of the discipline of ecology as a restorative process and healing art.[2]

Orr emphasizes the direct, participatory dialectic of the learning process, stressing that "environmental education ought to change the way people live, not just how they talk."[3] This is accomplished through the integration of school and community, through breaking the boundaries of disciplinary knowledge, and by involvement in practical, relevant projects that deal with the ecological relationships in a community. Orr's work at Meadowbrook, an intentional sustainable community, inspired the campus ecology movement, in which students have attempted to apply the theory of sustainability to resource use on college campuses, exploring ways to apply ecological principles practically to campus life.

Ecological literacy conveys an attitude, "driven by the sense of wonder, the sheer delight in being alive in a mysterious, beautiful world."[4] It promotes the importance of place, the ability to observe a landscape and to read its ecological nuances, the intimate knowledge of local ecology, a kinship for life, an aesthetic appreciation. These qualities are derived from familiarity, observation, study, open-mindedness, concern, and love.

Finally, Orr considers the relationship between place and pedagogy as crucial to ecological literacy. Citing John Dewey and Lewis Mumford, both of whom promoted the idea of a place as an educational tool, Orr suggests that "reeducating people in the art of living where they are" requires detailed knowledge of a place, the capacity for observation, and a sense of care and rootedness. Similar to Daniel Kemmis (see chapter 3), Orr regards place as the "bedrock of stable community and neighborhood."[5] He is primarily concerned with educational approaches to learning about place, as a means to understand ecological design, community building, and affiliation with the landscape. Place is the learning laboratory of ecological literacy. In the final section of this chapter, I describe the "sense-of-place map" as a culminating approach to ecological identity work.

Orr's book is an educator's guide to the requirements, applications, and implementation of ecological literacy, relating the idea to politics, community, schooling, and lifestyle. The idea of ecological identity is complementary and parallel to ecological literacy. As Orr suggests, ecological literacy requires the "more demanding capacity to observe nature with insight, a merger of landscape and mindscape."[6] Where ecological literacy creates an educational argument for teaching sustainability, ecological identity develops psychospiritual support for sustainable communities and psyches, with specific learning activities to support both goals.

These learning activities are not just for classrooms and structured educational environments. As Orr indicates, it is the tangible, everyday circumstances of living in a place that can serve as the best learning laboratory. In the next section, I describe ecological identity work as a guide to everyday action, as the foundation of reflective ecological practice. An important educational challenge is to integrate all aspects of our activities—linking professional practice, personal growth, and community participation. Reflective learning can take place anywhere, anyhow, anytime.

Ecological Identity and Everyday Life

The wilderness, the nature center, the highway, and the supermarket are all places in which we can see ourselves in a relationship to nature.

We assume that we go to the wilderness to enjoy solitude and to contemplate life, that we go to a nature center to learn about natural history, that we drive on the highway to get from one place to the other, that we shop in a supermarket to get food. Yet these settings are also learning environments—places to engage in ecological identity work.

For example, there is much more to shopping than typically meets the eye. Several years ago, I entered a supermarket with a Tanzanian colleague who had never been to such a market before. He looked at each item on the shelf as though he were examining a strange artifact. He studied the packaging, the materials, the advertising. He found the items on those shelves so interesting that he became immobilized, unable to move down the aisle. For him, this megasupermarket was a giant museum. There was so much to see and learn that he could only cover one small wing of the store without being overwhelmed. He couldn't possibly "shop" there unless I taught him how to do so. He had never seen so much stuff under one roof.

My friend was merely contemplating what I typically ignore. When I enter this supermarket I do so out of habit, and despite my best intentions, I become an habitual shopper. I am easily distracted by sale items or appetizing foods. I know that there is a politically correct way to shop. I should be aware of the amount of packaging that is used, whether the product is grown locally, whether it is "organic," which corporations are most exploitive of the natural environment, which foods use highly saturated tropical oils, and so on. But my Tanzanian friend showed me that my habits run even deeper. I take for granted the extraordinary wealth and security of places like this. His "beginner's eyes" were filled with a different kind of wisdom, allowing me to understand the extent to which I take material wealth and security for granted. Yet all the products on the supermarket shelves have a long and complex story, inevitably tied to questions of sustainable resources. Through this encounter, it occurred to me that ecological identity work can occur even in a supermarket.

What happens when we question our ordinary perception of driving a car? Most mornings I commute to work. I have a relatively short drive, about 25 minutes on a two-lane country highway that has light-to-moderate traffic. It is a pleasant drive through a hilly landscape, with several lakes and reservoirs and two picturesque small towns. I enjoy the ride because I have private space without interruption, a time when I can be alone, when the automobile insulates me and allows me to be in my own world. But rarely do I actually think about what I'm doing, that is, driving a car. When I shift to a more meditative mode, I realize that I am moving through a landscape. Here I am, driving across the hills, in a mass of fast-moving metal, spewing pollution into the atmosphere, one of millions of cars roaming the world this morning.

As I am driving the car there are millions of places my mind might go. Even in the confined space of an automobile, I can connect to various electronic information pathways, a mere click of the on/off switch places me in a complex communications network giving me the news, the latest events, the sports, the weather, a variety of incessant chatter, some of it interesting, most of it inane, most of it serving as a distraction, a seduction, a way to avoid being with myself. Or I can dwell in my internal distractions, yield to my busy mind, let various worries enter, project scenarios, immerse myself in recursive loops of what-ifs, or dealing with various irritations and discomforts, wondering about other people's motivations, being in several places simultaneously. This car ride can wind up being an adventure in consciousness, especially if I take the time to be aware of what I am thinking about and trace my thoughts as they move from place to place, from idea to idea, from feeling to feeling. I might arrive at my destination without actually having seen anything, not knowing what type of clouds were in the sky, whether there were any migrating hawks, not connecting with the earth or considering the significance of driving my car through the landscape. By attending to ecological identity, I interpret the act of driving from a very different perspective, breaking out of my habitual thought routines.

When I am working hard I find it crucial to take long walks, so I can clear my head, find empty spaces, and dwell in the present moment. It is difficult to achieve this state of mind, even when I use various meditative techniques. Often I walk in the woods and for most of the walk I debrief whatever happened during the day, or what I expect will happen the next day. The woods serve as wonderful scenery, but they become merely a backdrop for my busy mind. Of course there is nothing "wrong" with taking walks to process the days events. But I can so easily be overwhelmed by ordinary distractions that I may never notice the trees in the forest.

Thoreau refers to this dilemma in his "Walking" essay.

> Of course it is of no use to direct our steps to the woods, if they do not carry us thither. I am alarmed when it happens that I have walked a mile into the woods bodily, without getting there in spirit. In my afternoon walk I would

fain forget all my morning occupations and my obligations to society. But it sometimes happens that I cannot easily shake off the village. The thought of some work will run in my head and I am not where my body is—I am out of my senses. In my walks I would fain return to my senses. What business have I in the woods, if I am thinking of something out of the woods?[7]

Thoreau chastises himself for being unable to let go of whatever is on his mind. The woods are a sacred realm, and when he is in the woods, he expects to be there fully; in mind, body, and spirit; aware of the flora and fauna, his animal self; taking it all in, integrated with the natural world. He recognizes just how hard it is to accomplish this presence of mind. But what really happens when Thoreau, or for that matter anyone, is walking through the woods? What exactly do we see? There are just as many possibilities and choices in the woods as there are in the city and we can be just as easily overwhelmed by the infinity of perceptual vistas.

We can't experience ecological identity just by saying, "I think I'll spend some time in the woods today." The woods may not be nearby, or we might not be in the appropriate frame of mind. Ecological identity work requires the ability to overcome both internal and external distractions, achieving a state of mind, a way of being, an approach to life experience, and a philosophy of learning. The challenge is to experience ecological identity everywhere, not just in specific places—contained regions such as nature centers or parks—but in the various domains of everyday life. Often we reserve a time of the day or several weeks out of the year to schedule our time to be in nature. But we are really in nature all of the time.

The great risk is that ecological identity work becomes compartmentalized, fragmented, dislocated, something that we squeeze into "reserved" moments," just another activity on the weekly planner, or something that people discuss in environmental studies courses. Perhaps we view our time in nature as special and sacred and the rest of our activities as distractions. However, ecological identity work serves to integrate our lives; it enables us to envision ourselves as human beings in nature, to be aware of the habits and distractions that promote forgetfulness, to allow us to be attentive and present. It is a way of thinking about our connections to the earth—the places where we live, the walks we take, the water we drink, the food we eat, the air we breathe, the various ways we consume natural resources.

As a curriculum for everyday life, ecological identity is oriented around four overriding questions:

- Where do the things that I consume come from?
- What do I know about the place where I live?
- How am I connected to the earth and other living beings?
- What is my purpose and responsibility as a human being?

These questions are the foundations of environmental education. They are so fundamental to human existence that we usually take them for granted, yet they are a reminder of our forgetfulness, our inability to focus on what is at hand and what is important. Most of us think about these questions only sporadically, when a crisis compels us to look at things differently. When there is a power blackout, we think about where energy comes from. When there is a water shortage, we think about aquifers and water cycles. When there is death or loss, we think about the meaning of life. Ecological identity work brings these questions to our full attention, by helping us to consider them as integral to our daily awareness—as practitioners or citizens, in a variety of roles, and in unlimited settings.

Ecological Identity and Educational Design

When I taught a course on ecological identity for students and faculty at Charles University in Prague, I assumed that the group would at least have some familiarity with reflective learning in a formal educational setting. I came prepared with my full portfolio of progressive teaching techniques. Yet when I asked people to break out into small groups or to use artwork to express themselves, they had no idea what to do (language barriers notwithstanding). Exclusively exposed to didactic pedagogical models, all they knew were lectures, essays, and tests. Reflective learning, especially applied to environmental studies was a novel, unprecedented approach.

I realized that my educational philosophy was so deeply ingrained and so specific to the types of students and practitioners I was accustomed to that I failed to anticipate how reflective learning could be perceived as radical pedagogy. For most of the workshop, I carefully explained the philosophy behind every activity and how ecological identity work would improve their professional capacities and their ability to learn about environmental issues. Happily, through patience, perseverance, and the cross-cultural relevance of the activities, the approach caught on, and I ended up with an active, participatory group.

In this section I outline some of the teaching techniques that I emphasize throughout this book, describing three basic principles of educational design for reflective learning, and highlighting nine interpretive modalities that inspire ecological identity work. Although my teaching experience is primarily with adult learners, many of these modalities are as relevant for kindergarten as they are for graduate school. They are for the trail, the nature center, the public forum, as well as the university classroom.

When I plan a course, workshop, or any learning experience, I think first about the group that I am working with, and the kinds of experiences that would be most valuable for them. Then I consider my own interests and expertise, finding a balance between my experience and the expectations of the learners. Once the conceptual boundaries of the material emerge, i.e., the information that should be covered and exchanged, I design the reflective activities that will catalyze the material.

For ecological identity work, the first principle of educational design is to *highlight the importance of the learner's experience*. The teacher's task is to develop methodologies and approaches that elucidate, amplify, interpret, and synthesize these experiences, and to do so in a collaborative setting. This is accomplished by providing the learner with vibrant, creative, relevant, and reflective projects. In many cases, students think that the most valid knowledge comes from the authority of the instructor or the textbook. Not so. No matter how thrilling and clear a lecture may be, or how much people learn from a book, it is always the project itself that inculcates the deepest learning. Lectures and books are resources for the learner. The *project* represents the integration of theory and practice. That is why projects must be considered and designed carefully, geared toward the learner's experience, and be of direct personal and professional concern.

A learning community develops when there is an atmosphere of trust, inquiry, support, and common purpose. The classroom (in whatever form it takes) is a commons in which the participants contribute to a collaborative process, recognizing that their work is validated, supported, and critiqued in conjunction with their colleagues—understanding that the whole is far greater than the sum of its parts. The second principle of educational design is to *establish open, cooperative learning spaces*. Speaking and listening (see chapter 4) are fundamental learning community skills, the ability to hear what other people say, the space to express ideas freely, knowing what is appropriate and what contributes to group learning. The teacher sets an example by facilitating participation, respectful of the fact that some people are more willing to speak than others. A group requires a balanced discourse that is inclusive of various styles and temperaments. Ecological identity work can get deeply personal. It is the teacher's responsibility to respect the learner's privacy, to link individual experiences to a broader conceptual goal.

Third, the teacher must *provide conceptual vision,* keeping a group on track, focused, disciplined, and attentive to the broader purpose of the work at hand. Let the learning community take credit for its insights, build confidence among the participants, legitimate their knowledge and experience, empower them to interpret the material. The teacher knows when to let his or her own perspective emerge, attempting to synthesize diverse observations, speaking from personal experience when appropriate, providing a theoretical interpretation at times, but also knowing when to stay in the background, when to let the community move forward on its own. The teacher is an orchestrator, blending the talents of the learning community.

These principles are enhanced by attending to the *interpretive modalities* described below in the sections that follow.

Transforming the Ordinary

"If you are a poet, you will see clearly that there is a cloud floating in this sheet of paper. Without a cloud, there will be no rain; without rain, the trees cannot grow; and without trees, we cannot make paper. The cloud is essential for the paper to exist. If the cloud is not here, the sheet of paper cannot be here either. So we can say that the cloud and the paper inter-are. 'Interbeing' is a word that is not in the dictionary yet, but if we combine the prefix 'inter-' with the verb 'to be,' we have a new verb, inter-be."

Thich Nhat Hanh, *Peace is Every Step*, p. 5

What does it take to look at the world with beginner's eyes? When people face tumultuous life circumstances or encounter dramatic situations, they often reach new levels of awareness—a fresh perspective illuminates the ordinary. But how can this same perspective be generated through the habitual experiences of everyday life?

For example, if you are like most people, you know very little about the economic and ecological origins of the products you use every day. Consider something as mundane as a marking pen, the magic marker. To ask about the origins of a consumer item is to plunge into one of the great mysteries of contemporary life. What happens when we look deeply into the magic marker? If you have one around you, hold it up for a moment. Get a piece of paper and write down everything that you "know" about the marker. What is it made out of? How is it produced? What company makes it? What chemical processes are involved? What natural resources are used? What is the impact on the earth? Who owns the magic marker production company? How are the markers distributed? What happens to one when you are finished using it?

Most of us can figure out that it is an oil-based product, that the felt tip contains various dyes, that we buy the markers at a variety of retail outlets, that they probably wind up in a landfill somewhere. But it is unlikely that we know very much about the details of the production process, including the various power issues involved. Who owns the factory? Where is it located? What are the working conditions like there? Does the production process entail environmental pollution?

I use this approach as the basis of a teaching activity. I ask adult learners to tell me as much as they can about the magic marker. I divide a class into groups of four, giving each group several markers and a sheet of newsprint. Their task is to take 15 minutes to pool their knowledge and draw a diagram that traces the various resource transformation steps involved in the production, distribution, and consumption of the magic marker. They are free to organize the information however they choose. When they are finished I ask them to interpret their diagrams, commenting on the political and environmental implications of their observations.

Typically, the groups prepare idealized and symbolic charts. Their diagrams are colorful, humorous, and marginally factual. In a way they are mini–creation myths, imaginative stories based on limited information. For example, most groups trace the markers to the carboniferous age, aware that fossil fuels are a major component. They can fill in other broad outlines as well, but the information is nonspecific. What is most interesting are the series of questions they ask and the various interpretations they come up with. It becomes clear that the magic marker is merely a metaphor for a complex process that attests to the "magical" interpretation of natural resource transformation. We could substitute dozens of other commodities and arrive at the same result.

Thich Nhat Hanh, in *Peace is Every Step,* has a wonderful discussion about washing the dishes.[8] If you wash dishes just so you can move on to the next activity, or if you are resentful because you'd rather be doing something more creative, you are unlikely to learn much from dishwashing. But if you consider the deeper implications of dishwashing, whether it's attending more consciously to the actual dishwashing process, or by thinking about the wider ramifications of your actions—energy use, the watershed, food wastes, and so forth—then dishwashing becomes a reflective activity. If I ask a class to keep a journal about dishwashing as it relates to ecological identity and to read those journals at the next class session, then dishwashing becomes a focused learning activity.

Or if I send a group to the supermarket with a specific series of objectives (e.g., interview customers regarding their knowledge of the origins of various products, find out how many products are grown locally, study green marketing and packaging techniques), they will look at the supermarket very differently than if they were shopping for groceries. If the activity is handled well, they may never shop quite the same way again.

Transforming the ordinary requires imagination and opportunism—seizing the possibilities as they occur, considering the different ways you can shift perspectives. Consider the effects of doing without something (not using a personal computer or a telephone for a week), or dissecting an action (creating a property list), or doing something you don't ordinarily do (walking through a different part of town). For ecological identity work, it is particularly useful to focus on perceptions of place, observations of landscape, and habits of ecological familiarity. When these "shifts" are linked to broader learning goals, and are subjected to community discussion and personal reflection, they become modalities of interpretive insight.

The Collaborative Text

"Our culture is itself a vast writing space, a complex of symbolic structures. Just as we write our minds, we can say that we write the culture in which we live. And just as our culture is moving from the printed book to the computer, it is also in the final stages of the transition from a hierarchical social order to what we might call a 'network culture.'"

Jay David Bolter, *Writing Space*, p. 232

When a class is having a riveting discussion, I write their observations on a blackboard because they are generating a collaborative text, a series of insights that emerge from the learning community. Without this particular combination of people, these insights may never have developed. Their interpersonal chemistry and common environmental interests allow them to write an improvisational book, chapter by chapter, as the weeks of a class or the hours of a workshop float by.

During the first few weeks of a class, I design several "collaborative text" activities to give a group an appreciation of its collective wisdom. One approach is to assign a common reading (I often choose Thoreau's "Walking"). I ask students to select a short passage which they find particularly appealing, inspiring, or thought-provoking. Their task is to write the passage on a three-by-five index card. They must then attach a second card to the first one, explaining why they chose the passage.

In class, I hand out six additional index cards to each student. I ask them to hand me their original cards. After shuffling them I return them randomly. The students are then expected to read whatever passage they have received and attach additional comments, elaborating on the first two cards. Once again, the cards are collected, shuffled, and returned. The process is repeated until all of the index cards are used. At this point, everyone has a chain of cards, representing an anonymous and collective series of commentaries. I randomly choose several chains, and read them to the group.

Through this activity, the students plumb the depth of Thoreau's essay, but in a subtle and enlightening way. Under protection of anonymity, their responses to the passages are typically personal and profound. In effect, this process generates a collective wisdom, developing a shared document, reflecting their core environmental values. This exercise has a gentle bonding effect. What starts as a personal response and private dialogue becomes a group effort and a collective essay—the discourse of an intellectual commons. They lose their ego attachment to their personal comments, realizing that the strength of the text lies in their collective interpretation.

When students submit journals or essays, I invariably copy one or two passages from each essay and develop an anthology, derived from student work. I organize the passages according to the relevant themes of the class. The result is a short, collective essay, without names, reflecting the diverse opinions and insights of the group. Hence their individual essays become means to a collective end, an evolving, collaborative text. With hypertext computer applications, there are many logistical variations on this theme. Whatever the means, what is crucial is to allow the class to appreciate its own insights, without attaching them to individuals, thus facilitating a sense of community, a solid foundation for future work.

Multiple Learning Styles

"No longer is knowing confined to the symbol of the word; no longer is knowing confined by the processes of ratiocination. Thus, multiple modes of knowledge are possible in multiple domains. As we move away from the manipulation of linguistic and mathematical symbols, the knower increasingly becomes one of the defining characteristics of knowledge. Knowledge, as such, cannot be found in words, books, or computers, but only as it is animated by intelligence."

Jeffrey Kane, "On Knowing and Being," pp. 2–3

People have highly individualized learning styles—approaches to information, ways of interpreting experience, methods of conveying ideas, and so on, that vary according to how they learn and what is important to them. This is commonly referred to as multiple learning styles, an inclusive and diversified approach to education, indicating that there are many valid ways to learn something. One person would rather write an essay, another might like to develop a chart, another would prefer to draw a picture. Similarly, some prefer lectures, others like hands-on activities. The skilled teacher strives to achieve an appropriate balance. Ideally, teacher and learner alike should be proficient with several approaches, or at least be able to integrate them, developing a portfolio of learning styles. Versatility is not only a matter of the breadth of one's knowledge, it also reflects one's diversity of conceptual expression.

For ecological identity work, respecting multiple learning styles is crucial, as the goal is to find whatever means possible to convey and interpret one's experience of nature. In my teaching, I assign projects that allow the learner wide latitude in accessing the material, suggesting there are many paths to creative insight: essays, poetry, paintings, music, sculpture, dance, etc. The environmental tree (chapter 2) is an example of this type of project. For some people, an essay couldn't possibly convey their ideas, especially their psychospiritual understanding of the material, hence other forms of expression are more suitable. Others are intimidated by the prospect of using their artistic imagination, seeking shelter in their proficiency in, let's say, analytical writing. I suggest a middle ground, using projects not only to cultivate ecological identity but to develop multiple learning styles and to experiment with different forms of personal expression.

Adult learners are rarely asked to use artwork in educational settings, yet this opens new vistas of creativity and insight—unlocking important experiences, bridging conceptual boundaries, revealing theoretical insights, deriving practical applications. The use of artistic vision is a powerful teaching approach, not only because it encourages conceptual breakthroughs but for the sheer enjoyment of the process.

Also, the teacher must balance the textures of a class, understanding that a variety of teaching approaches enables students to thrive accordingly. Some students drift during lectures, but come alive in small group discussions, when they can take a more active role in the learning process. A good class can be like a magazine, with different sections and columns—essays, artwork, interviews, lectures, discussions. The teacher should construct a matrix of learning environments—respecting multiple intelligences, accessing diverse learning paths.

Museum Walls

"The arts should be positive, free, romantic, beautiful, something like a jewel, something which you cannot do without."

Hundertwasser, quoted in Harry Rand, *Hundertwasser*, p. 235

If you walk into a vibrant elementary school classroom, the walls will usually be covered with student work. This should be true for adult learners as well. When I assign a "multiple learning style" project, I ask my students to place their work around the room and invite them to spend as much time as they like interviewing one another, offering comments, and enjoying the work of their colleagues. This is another means of creating a learning community, as the participants gain a shared respect for the various talents of their peers. The environmental tree, the community awareness map, political autobiography—any project that has a visual component is ideally suited for public display.

In some cases, when space is available, the projects can be displayed in a common area, where people from other parts of the campus, other classes, or other departments can also view it. When a space is used temporarily, or students need to retrieve their work, then the display can be impermanent, a flash of creativity, enjoyed for the moment.

Autobiography

"Memory is the connections. Meaning comes from what something is connected to. Something unconnected, unassociated with, unrelated to anything is literally meaningless. Conversely something connected, associated, linked with many things is supercharged with meaning. And the farther back in time the connections go, the greater the meaning. By joining pieces of our lives together we create ourselves, free ourselves. It's all in the order and the sequence. For this reason, memory may be more in the way things are stored, rather than what is stored."

Lawrence Kushner, *The Book of Words,* p. 87

People are always fascinated to learn more about themselves and to have a constructive, focused way of doing so. Autobiographical analysis encourages learners to reconstruct their lives—coordinating their experiences into a coherent vision, linking the present to the past, providing continuity for the future. Ecological identity work is heavily based on the reconstruction of life experience, taking otherwise disparate memories and suggesting new ways to connect them.

Through autobiography, learners can experiment with their life stories, seeing how they fit with the experiences of others, understanding how their lives have a broader context. The purpose is not to dwell on details, wallow in successes or failures, or to provide forums for psychotherapy, rather it is to locate the patterns of experience, seminal moments when values are challenged, recalling people and incidents, understanding how personal development corresponds to ecological, historical, and political trends. The collective interpretation of autobiographical experience enables people to construct a common story as their lives become linked through their mutual interest in ecological or political identity.

Autobiographical projects serve as *developmental documents,* that is, interpretations that allow people to understand their values and ideals at a particular stage of life. Just as memorabilia from childhood or adolescence provide a glimpse into personal development, the adult learner can view ecological identity work in that sense. The environmental tree, the community awareness map, the sense-of-place map, political autobiography, these are chronicles of how people view themselves at particular stages of their lives. They are meant to be shared and saved. By sharing them, their full meaning is revealed, contributing to the public expression of ecological identity. By saving them, they serve as mileposts, benchmarks for evaluating personal change, guidelines for how to act and choose when difficult life choices are faced. I encourage my students to spread their lives before them, to use their experiences as a tapestry, projecting their aspirations into the future, signifying signposts or watersheds for their personal learning paths.

Constructive Controversy

"Groups that successfully use conflict for learning come to believe in themselves more strongly. With confidence in their ability to use conflict well, they can take more risks. Healthy conflict can get us more engaged in the problem-solving process—deepening our sense of ownership, both of the process and eventually, the solution."

Frances Moore Lappe and Paul Martin Dubois, *The Quickening of America*, p. 249

Although environmental issues are often controversial, when a relatively homogeneous group of students are involved in ecological identity work, it is easy for a subtle, insidious, politically correct consensus to emerge. People are unwilling to challenge one another because they don't want to upset anyone, or they feel that conflict may threaten the viability of the learning community. Many teachers avoid controversial issues, feeling that conflict may be anxiety-producing and potentially disruptive.

Yet in any group there is always much more diversity than meets the eye, and within seemingly consensual environmental groups (see chapter 2), there are many highly charged controversies. The teacher should cultivate this as controversy creates the "cognitive dissonance" that enables people to rethink their cherished assumptions. This requires a safe learning process in which people can express themselves without being stereotyped or ridiculed. The teacher has an important responsibility: to promote the minority viewpoint, to respect alternative perspectives, to legitimate opposing points of view, and to encourage respectful dissent.

From a design perspective, the teacher should consider the explicit and implicit controversies in the subject matter, developing assignments that allow contrasting opinions to emerge. Ultimately this contributes to the strength of the learning community—recognizing diversity and difference entails introspection, understanding the other, and searching for common ground. "Constructive" controversy implies that there is a protocol based on courtesy, cooperation, and mutua respect, knowing that one enters a controversy to better understand subject and to learn more about oneself.

Ecological Meditations

"It is too much to say that you are what you read in the morning, but i' sure bet that you aren't what you don't. A morning that begins with nun

words, or machines is likely to turn into a day filled with the same things. When a scanning of sky, trees, and birds begins the day, it could still turn out to be dominated by words and machines, but at least there would be a natural perspective to provide the larger context. A day that starts with a recognition of living processes can't be all bad."

Joseph W. Meeker, *Minding the Earth*, p. 3

There are many situations in which short, meditative experiences inspire creativity, affiliation, and community. This is most easily accomplished in the out-of-doors, when people can silently share a direct experience of nature. Given the various logistical constraints that may restrict bringing a group of students to a relatively wild natural area, the instructor should locate a proximate natural area of ecological interest that can serve as the site of a brief field trip. This can be a park, a stream, a vacant lot—whatever is accessible.

The group can explore this area by engaging in a series of ecological meditations. Numerous variants are possible, depending on the type of area visited and the predilections of the instructor and the students. The instructor should select and create a meditation that is in harmony with the place. What distinguishes ecological meditation is that it focuses on natural or ecological phenomena. This may include any obvious theme or pattern: habitats, weather, biogeography, raptors, etc. The instructor can experiment with various approaches, including solitary and collective meditation, silent walks, or any suitable approach.

It's important to let students know that you are not planning any psychological surprises and that none of the exercises will be threatening. Rather the focus will be on silent, meditative observations, using ecological themes. This disclaimer is helpful in heterogeneous groups which may include people who have not experienced meditation or have stereotyped it in a negative way. It is imperative that the instructor try to make people feel comfortable about this experience.

Finally, after a sequence of meditation and observation, the group should reconvene and discuss their experiences. As a debriefing technique, I ask the participants to describe any impressions they had. The critical step is to initiate a discussion about why we often deny ourselves these experiences and how we might overcome these blocks. I ask students to consider how to develop routines and rituals that will incorporate such meditative outdoor excursions into their daily life. Invariably, when I lead such excursions, people express their appreciation, as it opens up a new vista on the day, and legitimates their need to connect to the earth.

Mindfulness

"Mindfulness means paying attention in a particular way: on purpose, in the present moment, and non-judgmentally. This kind of attention nurtures greater awareness, clarity and acceptance of present-moment reality. It wakes us up to the fact that our lives unfold only in moments. If we are not fully present for many of those moments, we may not only miss what is most valuable in our lives but also fail to realize the richness and the depth of our possibilities for growth and transformation."

Jon Kabat-Zinn, *Wherever You Go, There You Are,* p. 4

One of my classes begins at 8:30 on Friday mornings. My students commute from all over New England, some from over 100 miles, driving 3 hours in their cars. It's the end of a week and as excited as they are to be in class, they are often in that peculiar zone of being in which they are tired and engaged at the same time. They have a lot on their minds. To focus their attention I ring a small bell and ask for a few moments of silence. This is a refreshing pause, allowing people to attend to the common purpose of their next few hours together—to be present in the learning community.

By bringing mindfulness to a learning experience you focus all of your energy and attention on the task of the moment, applying concentration and reflection. You empty your mind to watch it flow, quiet your thoughts to expand your thinking, feel your breath to awaken awareness. Meditation teachers describe mindfulness practice as the ultimate mind learning tool because it expands awareness through attending to the present.

I suggest to my students that if they experience creativity blocks, or get stuck in unproductive loops, or become overwhelmed by the magnitude of an assignment, that they take a few moments to practice mindfulness, and then start again. Or if a learning community begins to move too quickly, spinning through waves of ideas, enamored more with the pace than the pattern, swimming in rapidity and complexity—then it is time to pause and reflect, to pay attention to the learning path of the present moment.

Improvisational Excellence

"Our skills and works are but tiny reflections of the wild world that is innately and loosely orderly. There is nothing like stepping away from the road and heading into a new part of the watershed. Not for the sake of newness, but for the sense of coming home to our whole terrain. 'Off the trail' is another name for the Way, and sauntering off the trail is the practice of the wild. That is also where—paradoxically—we do our best work. But we need paths and trails and will always be maintaining them. You must first be on the path, before you can turn and walk into the wild."

Gary Snyder, *The Practice of the Wild,* p. 154

Teaching a class is like playing basketball or performing jazz. There is an organization and a structure, but the most exciting moments occur when the team or group discovers a zone of excellence, an improvisational communication that transcends the original structure. Everyone is clicking because they understand their common purpose and can meet whatever athletic or musical challenge they face. If the group adheres rigidly to a previously imposed structure, they may be unable to adapt to the changing circumstances.

When I go into a class with a fixed idea of what I want to accomplish and I myopically pursue my original goal without understanding the flow of the material or the interests of the students, I breed resentment or tedium. Sometimes a group takes a discussion or activity in an entirely unanticipated direction. The skillful teacher knows how to recognize this and will respond accordingly. In the middle of a class, when things are going well and group creativity is evident, someone will have an incredible idea for an activity or discussion. This is the time to take a chance and follow your intuition. See what path the class may follow.

The Sense-of-Place Map

When I first started teaching, I relied exclusively on lectures, discussions, and readings. And although those traditional methodologies have their place in education, I realized that for ecological identity work, something more dynamic was necessary, projects that would inspire my students to look deeply into their lives—modalities of reflective learning. In conceiving this new approach, I required something versatile, participatory, and substantive, a project that would integrate academic learning with personal experience, a studious rigor with artistic vision, an assignment that was simultaneously profound and accessible, intuitive and analytical.

It occurred to me that sense of place was literally the roots of ecological identity—ideas such as bioregionalism, sustainability, material simplicity, community, citizenship, decentralization, environmental psychology, and others were integrated in this one expression. All of my students and colleagues were cognitively, affectively, and spiritually motivated to understand and articulate their sense of place.

In 1977, as an experiment, I assigned a sense-of-place map as the final project for a class called History and Philosophy of Environmental Issues. It was simply an attempt to try a new approach to teaching. My philosophy of education was in its formative stages. I hadn't developed the language or rationale for understanding my work. The sense-of-place maps were exhilarating. The combination of creative energy, academic insight, and community bonding was inspiring. This was a watershed for my teaching career. I realized that this was the path I must follow.

Seventeen years later, after looking at hundreds of these sense-of-place maps, I realize that the project has become a rite of passage, not only for my class but as a celebration of reflective learning. It is the most direct example of the philosophy of education that inspired this book. It incorporates all of the components of this chapter: the domains of knowledge, ecological literacy, ecological identity and everyday life, educational design principles, and interpretive modalities. I describe the sense-of-place map as a culminating example of ecological identity work, an integrative synthesis of this chapter, and the concluding message of the book.

On the last day of my Patterns of Environmentalism class, the students display and describe their sense-of-place maps. These maps tell stories of ecological identity—how personal development reflects and is influenced by feelings and perceptions of landscape and habitat. The maps become community documents, providing closure to our common learning experience. They are trailmarkers on the converging paths of ecological identity work—blazing trails home (figure 6.1).

The sense-of-place map is an artistic representation of the ecological, geographical, developmental, and symbolic themes, patterns, and affinities of our lives. In compiling these maps, we consider the widest range of artistic mediums and expression: mobiles, mandalas, linear maps, surreal maps, dioramas, poetry, etc. We use the full range of our imagination in order to consider the deep meaning of the places where we live.

Sense of place is a search for ecological roots. This is best accomplished when we have a relationship to the land on which we live, when we can place ourselves securely in a tangible place. It is through the place where we live that we construct our personal identities, relate to the landscape, and determine what is important in our lives. Sense of place concerns our home and region, feelings about land and community, kindred species, community niches, and sacred places. To have a sense of place is to merge our personal geography with the ecological landscape, incorporate maps of memory with how we dwell in a bioregion.

The sense-of-place map is a rite of passage that links ecological identity to life-cycle development. What are the feelings, events, and choices that characterize how we see ourselves in nature through different periods of our lives, through various dwellings and travels in time and space? How might we communicate and illustrate the places we've been, where we live now, and where we see ourselves in the future? These questions provoke a process of self-reflection through which we perceive the places in which we live: the people, community, land, and species that form our networks of domesticity and exploration, the sources of sustenance and struggle. Sense of place is the domestic basis of environmentalism; it's the foundation of our deepest connection to the natural world. At the same time, we may observe how hard it is to establish meaningful community—how places are so easily eroded, how we dwell in so many mental spaces. From our past we recall special places that may no longer exist, but reside as internal sanctuaries from which we derive profound lessons.

Through this process we can trace our family, ethnic, spiritual, geographical, and ecological roots. These cognitive maps are forms of storytelling that provide a way to place our various journeys in perspective. What is the meaning of travel? What is significant about the places where we've been? What landscapes are we attracted to? In what kind of place would we like to live in the future? The sense-of-place map includes sacred places, places where we worship the earth, places that have special meaning, power, and significance.

It's about our habits of familiarity, the places we visit every day, how our immediate environment influences the way we think, breathe, and eat; the way the spirit of the land permeates our lives. It also reflects the ways we earn a living, the things we do to survive, the material basis of our life. In what ways are we bonded to the landscape? What are our emotional attachments to the place we live in? How do we coexist with other people, with the flora and fauna? How do we understand the local ecosystem? How do we define our bioregion?

Doug Aberley edited an excellent anthology, *Boundaries of Home,* in which he suggests the various ways that mapping can be used to help people understand what is important about the place they live in. It includes extensive sections on mapping the experience of place and a thorough primer on how to map a bioregion. The emphasis in the book is on mapping as a tool for community activists, but Aberley also recognizes the extraordinary learning potential of a map. He believes that when people create their own maps, they better understand the ecological and political dynamics of their bioregion. Hence they are empowered to participate in community action. In a lovely introductory passage, Aberley explores the lure of mapping and explains why maps are so powerful:

> If you gather a group of people together and ask them about maps you will always get a lively response. Like the universal fascination with moving water, or the dance of a fire's flame, maps hold some primal attraction for the human animal. For some, it is the memory of a treasure map followed in youth, or a scramble to a mountain vista etched forever in personal memory. For others, it is an almost magical chance to see what otherwise is hidden: the relationship of hill to forest to settlement to ruin. And, for yet others, maps unravel the mysteries of the present and future of place through the depiction of fixed and flowing energy layered in patterns of opportunity. Whether in our minds, or printed on paper, maps are powerful talismans that add form to our individual and social reality. They are models of the world—icons if you wish—for what our senses "see" through the filters of environment, culture, and experience."[9]

Aberley explains that mapping has become the province of specialists and although we have access to maps, we have lost our ability to "conceptualize, make and use images of place."[10] The point of his anthology is for bioregionalists to reclaim mapping so it becomes a tool for re-envisioning the landscape, rearranging power relationships, and reinhabiting place. The language of a map need not be technical. Rather mapping is about interpretation. "Reinhabitants will not only learn to put maps on paper, maps will also be sung, chanted, stitched and woven, told in stories, and danced across fire-lit skies."[11]

This approach reflects the spirit of the sense-of-place map activity. Maps are compiled according to the special skills and talents of the individual. The map becomes a form of storytelling, a document that explains a person's relationship to the earth. So the display and discussion of the map are as important as the compilation. As students design and implement their sense-of-place maps, they engage in a deeply reflective process that helps them visualize and represent their ecological identity. When they explain the maps in group settings, the maps become community documents.

In searching for our roots, we often experience how uprooted we really are. We realize the different number of places we have lived in. We see the various landscapes and traditions that have influenced us and wonder whether we have developed sufficient knowledge or intimacy in any of those realms. We recognize that we may be affiliated with many places simultaneously, extending time and space to live in regions of our own invention, carving an individualized flag of citizenship. How do we distinguish between these forces and pulls and recognize such multiple affiliations without sacrificing the intimacy of familiarity?

We may discover how difficult it is to delineate boundaries. These distinctions have been the cause of much global suffering as people have conflicts over land rights, ethnicity, and the haughty pride of extreme regional identification. Bioregionalism has a shadow side, the devolution into insular principalities, a distrust of those from outside the region. Gary Snyder calls nationalism "the grinning ghost of lost community."[12] Sense of place can be too provincial and overly idealized. Strong communities allow for transregional differences and a respect for diversity and difference.

Inevitably we must ask how the place we live in is connected to the global community. The various local/global clichés sound great, but they are very difficult to translate into action. What does it *mean* to act with the globe in mind? What are the boundaries of ecological practice? How do we begin to apprehend the global impact of our action, the global influence on our behaviors? A place is like a fractal. The more we explore it, the more we realize how the place expands beyond our limited perceptual sphere, how forms of communication take place within and between spaces, how our perception of space is framed by what enters our world.

Perhaps we need new terminology to understand the ecological and political meaning of traditional geographical space. In the twenty-first century, global citizens will form networks and allegiances based on *pluralistic regional identities*. That is, people will identify with many different places at once.

A language of regions should reflect not just geographical and political boundaries, but also the ecological and psychological aspects of place.

Bioregion is the idea that distinctions between regions should follow such factors as watersheds, landforms, cultural perceptions, spirit places, and elevation. These distinctions are based on ecological and cultural criteria rather than on patterns of land ownership. A set of multiple regions may be called a *transregion*. This idea helps us understand the connections between dynamic boundaries. Transregions allow us to explore how different habitats and cultures may trade, exchange information, and peacefully coexist. For example, the political integration of neighboring bioregions is a transregional issue. The *enviroregion* permeates all regional distinctions. It represents integrated global circulatory systems (ecological, cultural, spiritual). It describes the connecting processes that are the basis of global citizenship: the awareness that local bioregional actions, local states of mind, and local political decisions must always be informed by the enviroregional perspective. Finally, with the growing influence of electronic networks, we have new regions of communication spaces. These are postmodern spatial domains which allow people who live in bioregions that are not neighboring to develop networks and affinities. Mind regions that cut through bioregional distinctions are *metaregions*.

In the course of an ordinary day, we traverse all of these regions. Today's weather may have originated in the Gulf of Mexico or in the Canadian Arctic. Migrating songbirds range throughout mutliple habitats. It is not possible just to be concerned with the preservation of our special small corner of the globe. So our sense of place is as deep and extensive as we choose to make it. But this need not be the cause of confusion and dismay. Rather, it can teach us humility and respect. The world is far more diverse and complex than we can ever know. Local and global are merely convenient distinctions.

The sense-of-place map teaches us that we may travel in many directions, but our minds and hearts require roots. When we are rooted to the place where we live, it is easier to see the whole, to see ourselves as part of the landscape. When we care enough about life to learn about our place, we understand more about our neighbors. We create the potential to nurture compassion for all beings.

When I was 8 years old, I had an extensive map collection. Road maps were free in those days and whenever my family would go on a trip and stop for gas, I would race out of the car into the gas station and grab whatever maps I could find. I discovered all kinds of ways to study and play with the maps, inventing games and fantasies, organizing my collection, comparing the quality of the maps. I loved browsing through atlases, bathing in the colors and patterns, learning about where minerals came from, observing the subtleties and nuances of geographical relief. But I was most intrigued by local street maps because they allowed me to plan bicycle trips through the suburban developments of the south shore of Long Island. I would create my own maps, usually emphasizing the roads and highways, paying little attention to the homogenized landscape.

Years later I discovered a remarkable book about the area where I grew up. In *The Lord's Woods,* Robert Arbib described his own childhood adventures wandering through the same community. He grew up before the area was extensively developed. His adventures were through the streams and forests, looking at birds and frogs, playing in a more rural setting. Arbib bemoans the lost landscape, subverted by the scars of development, nostalgic for his childhood home. The book is about his fond memories of discovering the joys of his childhood landscape and his pain in realizing how suburban development had left so many emotional and ecological wounds.[9]

Wheras Arbib roamed in the so-called Lord's Woods, one generation later I roamed in the Lord's Suburban Development. I was desperate to learn more about the place where I lived. But no one taught me about its ecology, or about its history, or about the relationships that people had to the land. My parents were from Brooklyn. The suburbs added more green to their life. They lacked any awareness of ecological relationships. Who knew about such things in these 1950s suburbs? *Ecology* was an alien word to most American schoolchildren in the 1950s, and during my childhood I was searching for an ecological context for my life.

I remember when Hurricane Donna hit Long Island in 1960. Although I had some fear about the hurricane it was also one of the most beautiful memories of my childhood. The day after the hurricane the streets were flooded. Life as usual was disrupted. Fast-moving, billowy cumulus clouds moved across the sky. There was an eerie tropical light. It wasn't so much the disruption that I found so memorable. It was the beauty of nature, temporarily transfixing the landscape. We had no choice but to watch and listen to the weather. Snowstorms would have the same effect. I found that I might have a sense of place after all, but I would have to look at the sky to find it. So I developed an interest in the weather because it was the most dynamic and visible way for me to connect to the earth.

I found my sense of place in backyard baseball fields, on suburban roads, on road maps of distant places. I listened to broadcasts in foreign languages on an old shortwave radio I found in the attic. I walked on the beach and gazed at the ocean. I looked at astronomy books and learned about earth's position in the solar system. I dreamed about floating in New York Harbor in the days before European settlement, watching the forested landscape, swimming among the islands, roaming the coastline. I was searching for a personal geography, but I had no land to ground me. In fact, I lived on a landfill. Something, I knew, was missing.

I have chosen to raise my family in the rural, hilly, hardscrabble terrain of southwest New Hampshire. I have placed down new roots, making a commitment to the land and the community, trying to incorporate notions such as posterity and sustainability into my life. My kids can safely roam in the woods, yet often they choose to play inside. Still, they find nesting turtles, assorted scats, garnet-filled stones, glacial boulders. They are just as comfortable in front of the computer as they are on the trail. Their sense-of-place map is located in the twenty-first century. I wonder how they will recall the magic years of placemaking, how it will influence their lives, how they will bond with the earth. It takes a long time to learn about where you live. At least several generations. Will they have that kind of patience? This is what I hope they learn, that their lives will be sense-of-place maps, rooted in ecological awareness, that they will be cognizant of the flora and fauna, be respectful of their neighbors, show compassion for their community, that they will expand their horizons, negotiate metaregions and bioregions, travel many realms, but understand the profound significance of their home.

Epilogue: Ecological Identity Is a Way of Saying Grace

"To meditative minds the ineffable is cryptic, inarticulate: dots, marks of secret meaning, scattered hints, to be gathered, deciphered and formed into evidence; while in moments of insight the ineffable is a metaphor in a forgotten mother tongue."

Abraham Joshua Heschel, *I Asked for Wonder*

It's a cold morning in early February. The snow pack is covered with a thin sheet of ice, a remnant of last week's intense storm, a trough of tropical air in the midst of an icy winter. Frozen droplets are reflecting the sun, creating a spotted spectrum of color on the surface of the snow. The snow cover is a sea of waves and ripples. Frozen water! The trees are bare, stark, and bold, forming a network of infinite branches, a spare but dense thicket. There are no single trees, only the forest, a protective layer, one vast web of unfathomable interconnections.

I find an old tennis ball, and throw it out the window. I watch it roll down the hill, glide around icy bumps, move through the contours of the landscape, bounce off trees, and finally come to rest in a shallow, shady pocket. I take another ball and step outside. I throw it as far as I can. I wish I had a whole bucket of balls, and could roll each one of them, tracing their paths, noting their infinite variations, detecting the patterns, watching them all roll down the hill at the same time. I put on my boots and coat and walk into the woods to retrieve the two balls. The thin coat of ice won't support me and I plunge waist-deep into the snow. It's very difficult to walk through the forest today. Too icy for skis, too thin for snowshoes. I guess I'll have to stay right here.

It's a calm day. I find a spot on the snow where I can perch precariously, but without slipping through. I bask in the tentative February sunshine, feeling the heat of the sun and the cold of the snow. A gentle breeze moves through me. I inhale deeply and I gaze up at the sky.

Is this ecological identity work? Sometimes I think that if I just sit someplace long enough, and enter into a deep absorption, I will become a part of nature, and my ecological identity will be revealed. But that never happens. Sometimes I think that if I am patient and learn about nature, and observe the intricacies of the landscape, I will derive all kinds of lessons about life and I will find what I am looking for. But the landscape and the local ecology are just too complex and mysterious. Sometimes I think that if I analyze my ideas, and put them into order, organize them sequentially, and interpret them for meaning, that I will figure it all out, and my search will be complete. But who am I to invent meaning? What purpose do my interpretations serve, other than to give me a sense of temporary order in a complex universe?

In many ways, I am striving for something that I will never gain; a level of understanding and knowledge that enables me to grasp my purpose as a human being. But it really doesn't matter whether I figure this out. It is the process of thinking about the question that guides my experience. I could throw a million tennis balls out the window, and they would all follow different paths. And each path would be unique. Although I could eventually determine a statistical pattern that describes the probability of any ball following a specific path, I can never know for sure where the ball might bounce. There are too many variables and uncertainties. This reflects the fortuitous and overwhelming challenge of the postmodern age: how to find meaning in the world, given the proliferation of ambiguity and uncertainty. My purpose as a human being is simply to search for meaning. I must accept the inevitability of this challenge, just as I must accept the inevitability of my own death. But once I fully realize the temporality of meaning, I can accept the ambiguity of the search.

Here is the challenge and dilemma: How can I construct an ethical and moral foundation for my actions, if I also accept the temporality of interpretive meaning? What makes any worldview more or less acceptable than any other? What makes ecological identity work so virtuous? It is presumptuous for anyone to proclaim the truth, given the relativistic nature of human experience. But we can detect patterns and trends that despite their temporality might provide lasting insight.

It is very clear, for example, that the number of extinct species rises dramatically each year. What is the meaning of this pattern? Many conservationists believe that species diversity contributes to a robust and stable ecosystem and that over time, a decrease in biological diversity threatens the ecological foundation of human life. From this, I derive the following chain of meaning: (1) human life depends on biological diversity; (2) biological diversity requires the integrity of ecosystems; (3) ecosystems are typically disturbed by human intervention; (4) human intervention often takes the form of commercial development; (5) many humans don't recognize the ecological impact of their actions; (6) other people and I must be educated to fully understand these events; (7) this educational process entails an understanding of, and appreciation for, the preciousness of life; (8) such an appreciation can only occur through environmental education and spiritual growth. From this sequence, I can derive enough motivation and confidence to develop a way of looking at the world that integrates ecological knowledge and spiritual awareness. It is enough to lend structure and purpose to my daily actions, to affirm my career choice, to motivate me to write this book. It is enough to construct moral vision and purpose.

Yet there is another way to arrive at the same conclusion. As I sit in the cold snow and the warm sun, I am struck by wonder and awe; I am grateful for this moment of life, I am grateful for my awareness of this place, I am intensely fulfilled by my participation in nature and my deepening appreciation of life. The purpose of my life is to be awake in this moment. This moment is temporary and infinite. I find the deepest significance in its utter ephemerality. Right now, I am a human being, sitting on a piece of land, feeling my body, using my senses, exploring the world, and delighting in its ineffability. In each case, I derive temporary meaning, but lasting wisdom. I realize that I am not alone in these convictions. The quest for meaning represents the core of the earth's wisdom traditions—it is an inescapable human birthright, spanning diverse cultures through historical time.

Charlene Spretnak, in *States of Grace*, challenges us to explore the wisdom traditions as a means to a larger end, what she calls the "recovery of meaning," as the basis for an ecological postmodernism. Her book is about how we can reclaim the core teachings and practices of these traditions for the "well-being of the earth community." The wisdom traditions, in conjunction with new dimensions of ecological thought, can help illuminate the central issues of our time:

In the area of mind, perception, mental suffering, and the cessation of mental suffering, who has gone further than the teachings of the Buddha? In the area of perceiving an intimate connection with the rest of the natural world, who has gone further than the spiritual practice of native people? In the area of consciousness of the body as intricately embedded in a relational web, who has gone further in ritual honoring of the Earthbody and our personal bodies than the contemporary renewal of Goddess spirituality? In the area of social ethics as an expression of our comprehension of the divine oneness, who has gone further in the West than the core teachings of the Semitic religions: Judaism, Christianity, and Islam?[1]

Spretnak considers the daunting task of ecological and spiritual reconstruction. She finds that the notion of grace is crucial to these efforts. She is concerned that the experience of grace is so difficult to achieve because we are so imprisoned by the trivialities of consumerism, by the reductionist ideologies of modernism and institutional religions, by the magnitude of the state, by the engines of myopic economic growth. It is through the recovery of what she calls the "ecological imperative" that we can break through these restrictions. In a beautiful passage, she explores how our sense of ecological wholeness can transcend our perceptual limits and integrate many different approaches:

Once we no longer feel like tourists in the natural world, more and more of the intense vitality and intricate interrelatedness of the sacred whole is revealed to us. Its wonders, if we have sloughed off at least some of our implanted fear, evoke celebration, hence the flowering in recent years of personal and communal rituals of gratitude and joy. Gaian spirituality is non-sectarian, sprouting up in Hebrew tree-planting ceremonies, bioregional liturgies, the United Nations' interfaith Environmental Sabbath, the imaginative burst of homegrown solstice and equinox celebrations, and much more. Alas, disapproving glances still shoot between some members of these groups, as if there could be only one correct way to attempt to express the ineffable. Were we and the rest of the world set here by Raven the creator? By the Great Holy, the creativity of the cosmos? Were we modeled by God the Father? Did we grow from the body of Mother Earth? We are here—inextricably linked at the molecular level to every other manifestation of the great unfolding. We are descendants of the fireball. We are pilgrims on the Earth, glimpsing the oneness of the sacred whole, knowing Gaia, knowing grace.[2]

Gary Snyder concludes *The Practice of the Wild* with a statement about grace, explaining how grace is the first and last practice of the wild. Through the process of eating, we can reflect on ourselves as animals: how there is "no death that is not somebody's food, no life that is not somebody's death."[3] We must learn how to say grace; how to invent forms of grace for our own circumstances, how to add the saying of grace to our everyday lives. This is a tangible way to connect to the earth, the human body, our ancestors, and the full mystery of life and death. Saying grace is a reminder that there is an ethical dimension to our actions.

Ecological identity work is a way of saying grace. That is why it must always be grounded with the four basic questions that are the foundations of reflective environmental practice: Where do things come from? What do I know about the place where I live? How am I connected to the Earth? What is my purpose and responsibility as a human being? We consider these questions through the real circumstances of everyday life: domestic responsibilities, jobs, neighbors, family and children, home repairs, schools, and backyards. When we experience the significance of these moments, then our ecological identity exists in our homes and communities. Out of this shared wisdom, we discover how we connect to the earth, and what we have in common as members of the human community.

As we cultivate ecological identity, we become increasingly aware of the tensions, contradictions, and distractions that pervade our lives. We are no longer satisfied to live in forgetfulness and denial. We realize the necessity of balancing hope and despair, liberation and suffering, reflection and engagement. We learn how to find nature everywhere—how to see the ecological, political, and spiritual significance of everyday life. We will continue to be challenged by the shifting terrain of our cultural milieu. There is no escape from this. It is the reality of our times, the landscape of our lives.

As the world careens toward the twenty-first century, through ecological identity work we will continue to transform ourselves, inventing new metaphors and concepts, discovering rituals and practices, always asking ourselves how we should live. We shouldn't take too much credit for our accomplishments or be too harsh with ourselves over our mistakes. Our ideas are seeds, our language a species, our practice a flower. Let us find niches of responsibility, places of action, and spaces of meaning. Let us walk through our bioregions knowing that each step is significant and every breath a gift.

Notes

Chapter 1

1. C.T. Onions, *The Oxford Dictionary of English Etymology* (New York: Oxford University Press, 1991), 459–460. Weinstein and Platt in *Psychoanalytic Sociology*, 69, describe identity as:

> the complex interplay between bodily, psychological, and social processes which reaches a crucial point at the end of adolescence, an interplay which is the source of conflict that must be resolved in a successful synthesis of the variety of personal "choices" the individual has in defining what he is and what he will be. One result of this synthesis must be a stable sense of self, characterized by a feeling of "sameness and continuity." . . . [T]he present importance of the concept and the widespread awareness of it have more specific implications—namely, the existence of a society in which multiple roles, open-ended choices, and personal responsibility for choices are not only possible and permissible but obligatory.

In this sense, ecological identity represents the choice to formulate a personality and an approach to life based on a person's experiences in nature.

2. Edward O. Wilson, 396.

3. For an excellent discussion of nature as a "social construction," see Evernden.

4. Borden, 1. Also see Gray et al.

5 Borden, 4.

6. See Irwin Altman and Setha M. Low, *Place Attachment* (New York: Plenum, 1992).

7. The best discussion I have seen of the concept ecological worldview is in the doctoral dissertation of Malachy Shaw-Jones. In his research, Shaw-Jones interviews ten environmental studies students (from one of my classes) as a means to constructing a collective narrative, comprising the values that lead to an ecological worldview. From Shaw-Jones's perspective, the key elements of an ecological worldview are belonging and community (enlarged relational fields, identifications, biocentrism, global perspectives), compassionate values (empathy, inherent value of nature, pro-social outlook, ecological conscience), and the transpersonal (sacredness and mystery of nature, transcendental experiences). He suggests that the fruits of an ecological worldview are a synthesis of working out (behavior in the world), working in (personal development and change), and knowing (epistemology—head and heart).

Also see Goldsmith for a series of short essays exploring the metaphorical and scientific implications of ecological thought.

8. For an interesting perspective on the collective stories that constitute ecological identity, see the anthology *Story Earth,* compiled by the Inter Press Service. This is a collection of short essays and speeches by representatives of indigenous cultures. Most of the authors describe their relationship to the land and how it is threatened by global economic development.

9. Snyder, 26–27.

10. I am referring specifically to the following works: Cobb, *The Ecology of Imagination in Childhood;* Pearce, *Magical Child;* Shepard, *Nature and Madness;* and Hart, *Children's Experience of Place.*

11. Sobel, 159–60.

12. See Nabhan and Trimble, *The Geography of Childhood,* which is interesting from several perspectives: how the authors describe their childhood experiences, how they watch children in wild places, and how they tie together some of the environmental education literature on childhood experiences of nature.

13. The magazine *Orion* (winter, 1994) published a special issue on the ecology of love and loss. In addition to Windle's article also see in the same issue David Orr, "Love it or Lose it"; Jane Goodall, "Digging up the Roots"; John Elder, "The Turtle in the Leaves"; and Bruce Berger, "Comfort That Does Not Comprehend."

14. In *Silent Spring* Rachel Carson alerted the nation to the devastating impact of pesticides. This was in stark contrast to the passages of joyous reverie in her other books. The other books I refer to are Williams, *An Unspoken Hunger;* Kaza, *The Attentive Heart;* and Lopez, *Crossing Open Ground.* Also see Anderson, *Sisters of the Earth.*

15. Gottlieb, 207–212.

16. Leff, 282–283.

17. Leff, 284.

18. Leff, 285.

19. Rothenberg, quoted in Naess, 3.

20. Naess, 174.

21. Macy, 183. The chapter "The Greening of the Self" is most useful.

22. Roszak, 304.

23. Roszak, 320.

Chapter 2

1. Snyder, *The Practice of the Wild.* Chapter One, "The Etiquette of Freedom," contains a beautiful discussion of the idea of the wild.

2. Muir, 51–52.

3. Thoreau, "Walking," 612–613.

4. Thoreau, *Walden*, 364–365

5. For a comprehensive discussion of the wilderness philosophies of Muir and Thoreau, see Oelschlaeger.

6. Rachel Carson, *The Edge of the Sea*, quoted in Brooks, 169.

7. Thoreau, *Walden*, 343.

8. Worster provides an excellent discussion of Thoreau as a romantic ecologist. The section on "indoor science and Indian wisdom" is found on pp. 96–97.

9. All three quotes are from Thoreau, "Walking." "I think that I cannot," 594–595; "Moreover, you must walk," 596; "An absolutely new prospect," 598.

10. "The inseparability of life," Turner, 145.

11. For a historical perspective on the role of the wilderness journey, see Nash.

12. Edward O. Wilson's *The Diversity of Life* is an excellent example of how to integrate ecological theory, natural history, and romantic nature writing.

13. David Shi's, *The Simple Life* is a history of the idea of simple living. Both Thoreau and Muir are covered.

14. See Turner, *Rediscovering America;* Fox, *The American Conservation Movement;* Cohen, *The Pathless Way.*

15. Brooks, 228.

16. Consider Rachel Carson's passage about love and loss from *Silent Spring,* quoted in Brooks, 281.

I know well a stretch of road where nature's own landscaping has provided a border of alder, viburnum, sweet fern, and juniper with seasonally changing accents of bright flowers, or of fruits hanging in jeweled clusters in the fall. The road had no heavy load of traffic to support; there were few sharp curves or intersections where brush could obstruct the driver's vision. But the sprayers took over and the miles along that road became something to be traversed quickly, a sight to be endured with one's mind closed to thoughts of the sterile and hideous world we are letting our technicians make. But here and there authority had somehow faltered and by an unaccountable oversight there were oases of beauty in the midst of austere and regimented control—oases that made the desecration of the greater part of the road the more unbearable. In such places my spirit lifted to the sight of the drifts of white clover or the clouds of purple vetch with here and there the flaming cup of a wood lily.

17. For an environmental perspective on the history of American resource use, see Petulla.

18. For a discussion of conservation policy in Theodore Roosevelt's administration, see Samuel P. Hays, *Conservation and the Gospel of Efficiency: The Progressive Conservation Movement, 1890-1920* (New York: Atheneum, 1972).

19. McPhee, 67.

20. Worster's book traces the relationship between ecology and environmental thought. Fox provides an interesting overview of the history of mainstream environmental organizations, weaving his analysis around the archetypal example of John Muir. Borelli provides a discussion of the contemporary environmental spectrum, covering a wide span of philosophies and approaches. The first four chapters, in particular, cover the work of

contemporary environmental organizations and where they lie in the philosophical spectrum. Gottlieb's *Forcing the Spring* is an excellent history of environmental activism. For histories of the idea of wilderness, see Nash, and Oeschlaeger. The journalist Philip Shabecoff has written a popular history, *A Fierce Green Fire* and McCormick writes about the history of global environmentalism. Also see Hays, *Beauty, Health and Permanence: Environmental Politics in the United State, 1955-1985.*

21. Gottlieb, 10.

22. Gottlieb, 162–170.

23. Gottlieb, 9.

24. Edward O. Wilson, 243–281.

25. For a whimsical but often revealing look at contemporary environmental organizations, see Martel et al.

26. See Simmons for a comprehensive introduction to the various ways that environmentalism has influenced the construction of knowledge.

27. For Spretnak's "Ecofeminism: Our Roots and Flowering," see Diamond and Orenstein. The paths to ecofeminism are described on pp. 4–7.

28. Diamond and Orenstein, 5.

29. Examples of this research include Chodorow, *The Reproduction of Mothering;* Gilligan, *In a Different Voice;* and Conn, "Protest and Thrive."

30. On the role of ritual in environmental education, see Cynthia Thomashow.

31. See Oelschlaeger, 301–309, for a discussion of deep ecology, including a useful critique and plentiful references to the deep ecology literature.

32. See the now classic article by Naess, "The Shallow and the Deep, Long-Range Ecology Movements." Also see Warwick Fox's superb history of deep ecology, *Toward a Transpersonal Ecology*. Fox also deals with the numerous critiques of deep ecology and provides an excellent bibliography of the deep ecology literature.

33. The Institute of Deep Ecology Education is described somewhat in Devall.

34. The various branches of radical environmentalism can be daunting. Eckersley exhaustively analyzes the various ecophilosophical approaches, specifically in terms of their political relevance. In scanning the index, I noted the following conceptual categories, each representing a different philosophical orientation: animal liberation, anthropocentrism, autopoiesis, ecocentrism, ecocommunalism, ecofeminism, ecomonasticism, ecosocialism, emancipatory theory, human welfare ecology, humanist eco-Marxism, social ecology, and transpersonal ecology!

35. *Coevolution Quarterly 32*, winter 1981, 1.

36. Dodge, 6–7.

37. Dodge, 7.

38. Lewis Mumford deals with issues of scale, sustainability, and bioregionalism in many of his works, although he doesn't use those terms. His vision of an integrated, neighborhood-oriented, organic city, and his ideas about how to integrate the city with the countryside are very important to the bioregional perspective. In particular, see

Mumford, *The City in History.* E.F. Schumacher's *Small Is Beautiful* was a slogan of the 1970s. Leopold Kohr writes about the politics of scale in *The Breakdown of Nations.*

39. Snyder, *The Practice of the Wild*, 38.

Chapter 3

1. For a comprehensive discussion of common property resources within the context of ecological economics, see Costanza. Also see Berkes.

2. For an interesting historical perspective on image and identity, see Ewen.

3. On the commoditization of personal needs, see William Leiss, *The Limits to Satisfaction: An Essay on the Problem of Needs and Commodities* (Toronto: University of Toronto Press, 1976).

4. Weinstein, 89. Weinstein talks about property, money, and leadership as "transitional objects" and why they are so necessary to maintain a sense of personal and collective cohesion.

5. The environmental literature (not to mention that of political science and economics) on property rights is enormous. I recommend Mark Sagoff's *The Economy of the Earth* as an accessible and rigorous starting point.

6. On the commodification of land, see Polanyi.

7. Barber, *Strong Democracy*, 73. Chapter 4, "The Psychological Frame: Apolitical Man" is an excellent discussion of the psychological and political ramifications of individualism.

8. Needleman, 43.

9. Needleman, 59

10. See Boyte for a discussion of how social mobility and the information age change community awareness and inform citizen politics.

11. Bellah et al., *Habits of the Heart*, 333. This book is particularly useful because it penetrates so deeply into the everyday life questions that people care about. For anyone interested in why most Americans are ambivalent about civic participation, this is essential reading.

12. See "Virginia's Electronic Village," *The New York Times*, Jan. 16, 1994, 9. Apparently Blacksburg, Virginia has an electronic community network, which allows people to "peruse the minutes of town council sessions, keep track of various meeting schedules, communicate with local government officials, and maybe even conduct referendums on key issues." Thus Blacksburg is becoming an "interactive " town.

13. See Gergen, 210–216, for a discussion of community.

14. Bellah et al., 71–75, for a discussion of lifestyle enclaves.

15. Daly and Cobb, 170.

16. Daly and Cobb, 172.

17. Kemmis, 7.

18. Kemmis, 75.

19. Kemmis, 79.

20. See Berry, *The Unsettling of America,* and Snyder, *The Practice of the Wild* for discussions about politics and place. Also see Sale, *Dwellers in the Land.* Robert E. Goodin, *Green Political Theory,* describes how decentralization, place, and scale are fundamental to the Green political platform.

21. Kemmis, 117.

22. Kemmis, 75.

23. Havel, 104.

24. Havel, 65.

25. Ostrom, 1. See, in particular, chapter 3, "Analyzing Long-Enduring, Self-Organized, and Self-Governed CPR's." Ostrom also considers how bureaucracies succeed or fail in managing the commons. This is an excellent book, filled with interesting cases, and of great interest to the environmental community.

26. For an excellent and comprehensive discussion of how various social choice mechanisms and political systems succeed or fail at solving environmental problems, see Dryzek. Dryzek develops criteria for ecosystem management and then analyzes whether various political processes are capable of incorporating them. He concludes that radical decentralization and "practical reason" are most suited to this task. For a discussion of the relationship between local and global environmental politics, see Lipschutz and Conca.

27. Havel, 16.

Chapter 4

1. Arendt, "Communicative Power," 64.

2. "The vocabulary of assertive mutuality includes such words as co-action, interconnection, sharing, mutuality, integration, collaboration, cooperation, synthesis, vulnerability, and interdependence, and such phrases as agency-in-community, giving and openness to others, self-assertion as opposed to self-imposition, and the capacity to act and implement as opposed to the ability to control others," Kreisberg, 86. Kreisberg has an excellent discussion of "power over" and "power with." His book is an important resource for educators who wish to empower their students. Also see Boyte, who applies relational power to citizen politics.

3. For an anthropological discussion regarding family structure and political behavior, see Todd.

4. Lakoff, 2.

5. Lakoff, 21–22.

6. Lakoff, 17.

7. Lakoff, 12.

8. This emphasis on speaking and listening is compatible with socially engaged Buddhism. See Thich Nhat Hanh; Jones; and Dass and Bush.

9. Barber, *Strong Democracy,* 175.

10. For futher discussion of the ideological ramifications of so-called issue advertisements, see Mitchell Thomashow, "Corporate Advertisements and Environmental Futures." This essay decodes the ideological content of numerous enviromental ads. It also provides some curriculum ideas for the college classroom.

11. In particular see Ophuls; Heilbroner; and Hardin and Baden. For a more recent discussion of coercion as a "social choice mechanism," see Dryzek.

12. For an excellent anthology on the new theories and processes of international environmental politics, see Lipschutz and Conca.

13. For more on controversial issues and education see Mitchell Thomashow, "The Virtues of Controversy."

14. I am influenced by Naess, *Ecology, Community and Lifestyle*, 130:

All our actions, and all our thoughts, even the most private, are politically relevant. If I use a clipped tea, some sugar, and some boiling water, and I drink the product, I am supporting the tea and sugar prices and more indirectly I interfere in the works and capital conditions of the tea and sugar plantations of the developing countries. In order to heat the water, I may have used wood or electricity or some other kind of energy, and then I take part in the great controversy concerning energy use. I may use water from a private source or a public source, and in either case I participate in a myriad of politically burning questions of water supply. I certainly have a political influence daily in innumerable ways.

15. "The clock, moreover, is a piece of power-machinery whose 'product' is seconds and minutes: by its essential nature it disassociated time from human events and helped create the belief in an independent world of mathematically measurable sequences: the special world of science." Mumford, *Technics and Civilization*, 15.

16. Barber, *Strong Democracy*, 152, 155.

17. Lappe and DuBois in *The Quickening of America* provide an outstanding handbook of citizen democracy. They furnish dozens of examples of successful community action. Most important, they engage the reader in a series of reflective activities and choices to facilitate political empowerment and citizenship. For the reader who is interested in the educational applications of political identity or who wants to place community politics in a broader context, this is a superb book.

18. Havel, 8.

19. Havel, 2.

Chapter 5

1. Thornton, "The State of the Environmentalists 1993," 2

2. Thornton, "Radical Confidence," 44.

3. Macy, *World as Lover*, 16.

4. Elder, 24.

5. The most compelling and thorough critique of progress and economic growth remains Mumford, *The Myth of the Machine*.

6. Fingarette, *The Self in Transformation.*

7. For a systematic, theoretical treatment of the displacement of indigenous cultures in the New England landscape, see also Cronon. An excellent discussion of the contact between Western culture and North American indigenous culture can be found in Turner, *Beyond Geography.*

8. Fingarette, 163.

9. Fingarette, 164.

10. Fingarette, 166.

11. For a detailed evaluation of this retreat, including many comments from the attendees about its relevance to their work, see Bush.

12. Thich Nhat Hanh, "Protecting the Environment," 6. See also Young.

13. For an excellent, readable discussion of the various ways that stress enters our lives, and for a comprehensive program outlining what we can do about it, see Jon Kabat Zinn, *Full Catastrophe Living: Using the Wisdom of the Body and Mind to Face Stress, Pain, and Illness* (New York: Delacorte Press, 1990).

14. Snow, xxxi–xxxii. Slowly, this is changing. Increasingly, the leadership of the larger environmental organizations comes from business schools or the law. Nevertheless, most environmental administrators are not trained in management, but wind up in such positions because they work for small- to moderate-sized organizations where one person plays multiple roles and wears numerous hats. See Snow's study for more details.

15. The classic text of reflective practice is Schon's *The Reflective Practitioner.*

16. See Snow, chapter 2, "Staff Leadership."

17. Tenzin Gyatso, the Fourteenth Dalai Lama, 3–4.

Chapter 6

1. Orr, *Ecological Literacy,* 90.

2. Orr, 91.

3. Orr, 91.

4. Orr, 86.

5. Orr, 130. See Orr's excellent chapter "Place and Pedagogy," 125–131.

6. Orr, 86.

7. Thoreau, "Walking," 597–598.

8. Thich Nhat Hanh, 26.

9. Aberley, 1.

10. Aberley, 1.

11. Aberley, 5.

12. Snyder, 43.

Epilogue

1. Spretnak, 23.
2. Spretnak, 112–113.
3. Snyder, *The Practice of the Wild*, 184.

Bibliography

Aberley, Doug, ed. *Boundaries of Home*. Gabriola Island, British Columbia: New Society Publishers, 1993.

Alexander, Christopher, Sera Ishikawa, Murray Silverstein, with Max Jacobson, Ingrid Fiksdahl-King, and Shlomo Angel. *A Pattern Language*. New York: Oxford University Press, 1977.

Anderson, Lorraine. *Sisters of the Earth*. New York: Random House, 1991.

Arbib, Robert. *The Lord's Woods: The Passing of an American Woodland*. New York: Norton, 1971.

Arendt, Hannah. *The Human Condition*. Chicago: University of Chicago Press, 1958.

Arendt, Hannah. "Communicative Power." In Power, Steven Lukes, ed. New York: New York University Press, 1986.

Barber, Benjamin R. *Strong Democracy: Participatory Politics for a New Age*. Berkeley: University of California Press, 1984.

Barber, Benjamin R. *An Aristocracy of Everyone: The Politics of Education and the Future of America*. New York: Ballantine Books, 1992.

Belenkey, Mary Field, Blythe McVicker Clinchy, Nancy Rule Goldberger, and Jill Mattuck Tarule. *Women's Ways of Knowing: The Development of Self, Voice, and Mind*. New York: Basic Books, 1986.

Bellah, Robert N., Richard Madsen, William M. Sullivan, Ann Swidler, and Steven M. Tipton. *Habits of the Heart: Individualism and Commitment in American Life*. New York: Harper & Row, 1985.

Bennett, John W. *The Ecological Transition: Cultural Anthropology and Human Adaptation*. Elmsford, N.Y.: Pergamon, 1976.

Berkes, Fikret, ed. *Common Property Resources*. London: Belhaven Press, 1989.

Berry, Wendell. *The Unsettling of America*. New York: Avon Books, 1977.

Bolter, Jay David. *Writing Space: The Computer, Hypertext and the History of Writing*. Hillsdale, N.J.: Erlbaum, 1991.

Borden, Richard. "Ecology and Identity." In *Proceedings of the First International Ecosystems-Colloquy*. Munich: Man and Space, 1986.

Borelli, Peter, ed. *Crossroads: Environmental Priorities for the Future.* Washington, D.C.: Island Press, 1988.

Boyte, Harry C. *Commonwealth: A Return to Citizen Politics.* New York: Macmillan, 1989.

Brooks, Paul. *The House of Life: Rachel Carson at Work.* Boston: Houghton Mifflin, 1989.

Bush, Mirabai. *The Spiritual Roots of Social Activism: Reclaiming Our Common Strength for the Common Good.* A report for the Nathan Cummings Foundation and the New World Foundation, New York, 1993.

Carson, Rachel. *The Silent Spring.* Boston: Houghton Mifflin, 1962.

Chase, Steve, ed. *Defending the Earth: A Dialogue between Murray Bookchin and Dave Foreman.* Boston: South End Press, 1991.

Chodorow, N. *The Reproduction of Mothering: Psychoanalysis and the Sociology of Gender.* Berkeley: University of California Press, 1978.

Cobb, Edith. *The Ecology of Imagination in Childhood.* Dallas: Spring Publications, 1993.

Cohen, Michael P. *The Pathless Way: John Muir and American Wilderness.* Madison: University of Wisconsin Press, 1984.

Conn, Sarah A. "Protest and Thrive: The Relationship Between Global Responsibility and Personal Empowerment." *New England Journal of Social Policy.* Spring 1990.

Costanza, Robert, ed., *Ecological Economics: The Science and Management of Sustainability.* New York: Columbia University Press, 1991.

Cronon, William. *Changes in the Land: Indians, Colonists and Ecology of New England.* New York: Hill & Wang, 1983.

Daly, Herman E., and John B. Cobb. *For the Common Good: Redirecting the Economy Toward Community, the Environment and a Sustainable Future.* Boston, Beacon Press, 1989.

Dass, Ram, and Mirabai Bush. *Compassion in Action: Setting Out on the Path of Service.* New York: Crown, 1992.

Deetz, Stanley A. *Democracy in an Age of Corporate Colonization: Developments in Communication and the Politics of Everyday Life.* Albany, NY: State University of New York Press, 1992.

Devall, Bill, and George Sessions. *Deep Ecology: Living as if Nature Mattered.* Salt Lake City: Gibbs M Smith, 1985.

Devall, Bill. "Applied Deep Ecology." *The Trumpeter*, volume 10, no. 4, fall 1993.

Diamond, Irene, and Gloria Feman Orenstein, eds. *Reweaving the World: The Emergence of Ecofeminism.* San Francisco: Sierra Club, 1990.

Dodge, Jim. "Living by Life: Some Bioregional Theory and Practice." *Coevolution Quarterly* 32, winter 1981.

Dryzek, John S. *Rational Ecology: Environment and Political Economy.* New York: Blackwell, 1987.

Earth Works Group. *50 Simple Things You Can Do to Save the Earth.* Berkeley, Calif.: Earthworks Press, 1989.

Eckersley, Robin. *Environmentalism and Political Theory: Toward an Ecocentric Approach.* Albany, NY: State University of New York Press, 1992.

Ehrlich, Paul. *The Population Bomb.* New York: Ballantine, 1968.

Elder, John C. "The Turtle in the Leaves," *Orion*, volume 13, no. 1, winter 1994.

Elshtain, Jean Bethke. *Meditations on Modern Political Thought: Masculine/Feminine Themes from Luther to Arendt.* University Park: University of Pennsylvania Press, 1992.

Evernden, Neil. *The Social Creation of Nature.* Baltimore: Johns Hopkins University Press, 1992.

Ewen, Stuart. *All Consuming Images: The Politics of Style in Contemporary Culture.* New York: Basic Books, 1988.

Falk, Richard. *Explorations at the Edge of Time: The Prospects for World Order.* Philadelphia: Temple University Press, 1992.

Fingarette, Herbert. *The Self in Transformation: Psychoanalysis, Philosophy and the Life of the Spirit.* New York: Harper, 1963.

Fox, Stephen. *The American Conservation Movement: John Muir and His Legacy.* Madison: University of Wisconsin Press, 1985.

Fox, Warwick. *Toward a Transpersonal Ecology: Developing New Foundations for Environmentalism.* Boston: Shambhala, 1990.

Gardner, Howard. *Frames of Mind: The Theory of Multiple Intelligences.* New York: Basic Books, 1983.

Gergen, Kenneth J. *The Saturated Self: Dilemmas of Identity in Contemporary Life.* New York: Basic Books, 1991.

Gilligan, Carol. *In a Different Voice: Psychological Theory and Women's Development.* Cambridge, Mass.: Harvard University Press, 1982.

Goldsmith, Edward. *The Way: An Ecological World-view.* Boston: Shambhala Publications, 1993.

Goldstein, Joseph, and Jack Kornfield. *Seeking the Heart of Wisdom: The Path of Insight Meditation.* Boston: Shambhala, 1987.

Goleman, Daniel. *Vital Lies, Simple Truths: The Psychology of Self-Deception.* New York: Simon & Schuster, 1985.

Goodin, Robert E. *Green Political Theory.* Cambridge, Mass.: Polity Press, 1992.

Gore, Al. *Earth in the Balance.* Boston: Houghton Mifflin, 1992.

Gray, D.B., R.J. Borden, and R. Weigel. *Ecological Beliefs and Behaviors: Assessment and Change.* Westport, Conn.: Greenwood, 1985.

Grumbine, R. Edward. *Ghost Bears: Exploring the Biodiversity Crisis.* Washington, D.C.: Island Press, 1992.

Gottlieb, Robert. *Forcing the Spring: The Transformation of the American Environmental Movement.* Washington, D.C.: Island Press, 1993.

Halifax, Joan. *The Fruitful Earth.* San Francisco: Harper, 1993.

Hampden-Turner, Charles. *Maps of the Mind: Charts and Concepts of the Mind and Its Labyrinths*. New York: Macmillan, 1982.

Hardin, Garrett. *Living Within Limits: Ecology, Economics and Population Taboos*. New York: Oxford University Press, 1993.

Hardin, Garrett, and John Baden, eds. *Managing the Commons*. San Francisco, Calif.: Freeman, 1977.

Hart, Roger. *Children's Experience of Place*. New York: Irvington, 1979.

Havel, Vaclav. *Summer Meditations*. New York: Knopf, 1992.

Hays, Samuel P. *Beauty, Health and Permanence: Environmental Politics in the United States, 1955-1985*. New York: Cambridge University Press, 1987.

Heilbroner, Robert L. *An Inquiry Into the Human Prospect: Updated and Reconsidered for the 1980's*. New York: Norton, 1980.

Henderson, Hazel. *Paradigms in Progress: Life Beyond Economics*. Indianapolis: Knowledge Systems, 1991.

Heschel, Abraham Joshua. *I Asked for Wonder: A Spiritual Anthology*. New York: Crossroads, 1990.

Inter Press Service, *Story Earth*. San Francisco: Mercury House, 1993.

Jones, Ken. *The Social Face of Buddhism: An Approach to Political and Social Activism*. Boston: Wisdom, 1989.

Kabat-Zinn, Jon. *Full Catastrophe Living*. New York: Delacorte Press, 1990.

Kabat-Zinn, Jon. *Wherever You Go, There You Are*. New York: Hyperion, 1994.

Kane, Jeffrey. "On Knowing and Being," *Holistic Education Review*, volume 7, no. 2, Summer 1994.

Kaza, Stephanie. *The Attentive Heart: Conversations with Trees*. New York: Ballantine, 1993.

Kegan, Robert. *The Evolving Self: Problem and Process in Human Development*. Cambridge, Mass.: Harvard University Press, 1982.

Kemmis, Daniel. *Community and the Politics of Place*. Norman: University of Oklahoma Press, 1990.

Kohr, Leopold. *The Breakdown of Nations*. New York: Dutton, 1957.

Kreisberg, Seth. *Transforming Power: Domination, Empowerment and Education*. Albany: State University of New York Press, 1992.

Kushner, Larry. *The Book of Words*. Woodstock, Vt.: Jewish Lights Publishing, 1993.

LaChapelle, Dolores. *Sacred Lands Sacred Sex: Rapture of the Deep*. Silverton, Colo.: Finn Hill Arts, 1988.

Lakoff, Robin Tolmach. *Talking Power: The Politics of Language*. New York: Harper Collins, 1990.

Lappe, Francis Moore, and Paul Dubois. *The Quickening of America*. San Francisco: Jossey-Bass, 1994.

Leff, Herbert L. *Experience, Environment, and Human Potentials.* New York: Oxford University Press, 1978.

Lopez, Barry. *Crossing Open Ground.* New York: Random House, 1989.

Levey, Joel. *The Fine Arts of Relaxation, Concentration and Meditation.* London, England: Wisdom, 1987.

Lipschutz, Ronnie D., and Ken Conca, eds. *The State and Social Power in Global Environmental Politics.* New York: Columbia University Press, 1993.

Macy, Joanna. *World as Lover, World as Self.* Berkeley, Calif.: Parallax Press, 1991.

Marcus, Clare Cooper. "Environmental Memories." In *Place Attachment,* Irwin Altman and Setha M. Low, eds. New York: Plenum, 1992.

Martel, Ned, Blan Holman, and the Editors. "Inside the Environmental Groups, 1994." *Outside,* March 1994.

Marx, Leo. *The Machine in the Garden.* New York: Oxford University Press, 1964.

McCormick, John. *Reclaiming Paradise: The Global Environmental Movement.* Bloomington: Indiana University Press, 1989.

McPhee, John. *Encounters with the Archdruid.* New York: Farrar, Straus, Giroux, 1990.

Meeker, Joseph W. *Minding the Earth: Thinly Disguised Essays on Human Ecology.* Alameda, Calif.: The Latham Foundation, 1988.

Merchant, Carolyn. *Ecological Revolutions: Nature, Gender and Science in New England.* Chapel Hill: University of North Carolina Press, 1989.

Merchant, Carolyn. *Radical Ecology.* New York: Routledge, 1992.

Meyrowitz, Joshua. *No Sense of Place: The Impact of Electronic Media on Social Behavior.* New York: Oxford University Press, 1985.

Muir, John. *The Mountains of California.* San Francisco: Sierra Club, 1988.

Mumford, Lewis. *The City in History: Its Origins, Its Transformations and Its Prospects.* New York: Harcourt, Brace & World, 1961.

Mumford, Lewis. *Technics and Civilization.* New York: Harcourt, Brace & World, 1963.

Mumford, Lewis. *The Myth of the Machine,* vol. 2: *The Pentagon of Power.* New York: Harcourt Brace Jovanovich, 1970.

Nabhan, Gary Paul, and Stephen Trimble. *The Geography of Childhood.* Boston: Beacon, 1994.

Naess, Arne. "The Shallow and the Deep, Long Range-Ecology Movements." *Inquiry 16.* Oslo, 1973.

Naess, Arne. *Ecology, Community and Lifestyle.* David Rothenberg, trans. New York: Cambridge University Press, 1989.

Nash, Roderick. *Wilderness and the American Mind,* ed. 3. New Haven, Conn.: Yale University Press, 1982.

Needleman, Jacob. *Money and the Meaning of Life.* New York: Doubleday, 1991.

Nhat Hanh, Thich. "Earth Gathas." In *Dharma Gaia: A Harvest of Essays in Buddhism and Ecology*. Allan Hunt Badiner, ed. Berkeley, Calif.: Parallax Press, 1990, 195–197.

Nhat Hanh, Thich. *Peace is Every Step: the Path of Mindfulness in Everyday Life*. New York: Bantam Books, 1991.

Nhat Hanh, Thich. *A Guide to Walking Meditation*. Nyack, N.Y.: Fellowship, 1985.

Nhat Hanh, Thich. *The Blooming of a Lotus: Guided Meditation Exercises for Healing and Transformation*. Boston: Beacon Press, 1993.

Nhat Hanh, Thich. "Protecting the Environment." *The Mindfulness Bell 7*, summer/fall, 1992.

Nhat Hanh, Thich. *Being Peace*. Berkeley, Calif.: Parallax Press, 1987.

Oeschlaeger, Max. *The Idea of Wilderness*. New Haven, Conn.: Yale University Press, 1985.

Ophuls, William. *Ecology and the Politics of Scarcity*. San Francisco: Freeman, 1977.

Orr, David. *Ecological Literacy: Education and the Transition to a Postmodern World*. Albany: State University of New York Press, 1992.

Orr, David. "Love it or Lose it: The Coming Biophilia Revolution." In *The Biophilia Hypothesis*, Stephen R. Kellert and Edward O. Wilson, eds. Washington, D.C.: Island Press, 1993.

Ostrom, Elinor. *Governing the Commons: The Evolution of Institutions for Collective Action*. New York: Cambridge University Press, 1990.

Pearce, Joseph Chilton. *Magical Child*. New York: Dutton, 1977.

Petulla, Joseph M. *American Environmental History*, ed. 2. Columbus, Ohio: Merrill, 1988.

Polanyi, Karl. *The Great Transformation*. Boston: Beacon, 1968.

Rand, Harry. *Hundertwasser*. Cologne: Benedikt Taschen, 1991.

Ridgeway, James. *Who Owns the Earth*. New York: Macmillan, 1980.

Rockefeller, Steven C., and John C. Elder. *Spirit and Nature: Why the Environment Is a Religious Issue*. Boston: Beacon, 1992.

Roszak, Theodore. *The Voice of the Earth*. New York: Simon & Schuster, 1992.

Rothenberg, David. *Hand's End: Technology and the Limits of Nature*. Berkeley, Calif.: University of California Press, 1993.

Sagoff, Mark. *The Economy of the Earth: Philosophy, Law and the Environment*. Cambridge, England: Cambridge University Press, 1989.

Sale, Kirkpatrick, *Dwellers in the Land: The Bioregional Vision*. San Francisco: Sierra Club, 1985.

Schon, Donald A. *The Reflective Practitioner: How Professionals Think in Action*. New York: Basic Books, 1983.

Schumacher, E.F. *Small is Beautiful: Economics as if People Mattered*. New York: Harper, 1973.

Shabecoff, Philip. *A Fierce Green Fire*. New York: Hill & Wang, 1993.

Shaw-Jones, Malachy. *Ecological Worldview*. Ann Arbor: UMI, 1992.

Shepard, Paul. *Nature and Madness*. San Francisco: Sierra Club, 1982.

Shi, David. *The Simple Life: Plain Living and High Thinking in American Culture*. New York: Oxford University Press, 1985.

Simmons, I.G. *Interpreting Nature: Cultural Constructions of the Environment*. New York: Routledge, 1993.

Snow, Donald. *Inside the Environmental Movement: Meeting the Leadership Challenge*. Washington, D.C.: Island Press, 1992.

Snyder, Gary. *Turtle Island*. New York: New Directions, 1974.

Snyder, Gary. *The Practice of the Wild*. San Francisco: North Point Press, 1990.

Sobel, David, *Children's Special Places: Exploring the Role of Forts, Dens, and Bush Houses in Middle Childhood*. Tucson: Zephyr Press, 1993.

Spretnak, Charlene. *States of Grace: The Recovery of Meaning in the Postmodern Age*. San Francisco: Harper, 1991.

Steier, Frederick, ed. *Research and Reflexivity*. Newbury Park, Calif.: Sage, 1992.

Tenzin Gyatso, the Fourteenth Dalai Lama. *Compassion and the Individual*. Boston: Wisdom, 1991.

Thomashow, Cynthia. "Seeking the Sacred in Everyday Life." *Holistic Education Review*, vol. 6, no. 3, Autumn 1993.

Thomashow, Mitchell, and Fred Taylor. "Voices of Environmental Identity." *Whole Terrain* vol. 2, 1993.

Thomashow, Mitchell, Jimmy Karlan, and David Sobel. *KNOW NUKES: Controversy in the Classroom*. Keene, N.H.: Antioch/New England Graduate School, 1985.

Thomashow, Mitchell. "Corporate Advertisements and Environmental Futures." *Bulletin of Science, Technology and Society*, vol. 8, no. 1, 1988.

Thomashow, Mitchell. "The Virtues of Controversy." *Bulletin of Science, Technology and Society*, vol. 9, no. 1, 1989.

Thoreau, Henry David. "Walking." In *The Portable Thoreau*, Carl Bode, ed. New York: Penguin, 1979, 592–630.

Thoreau, Henry David. *Walden*. In *The Portable Thoreau*, Carl Bode, ed. New York: Penguin, 1979, 258–572.

Thoreau, Henry David. "Life Without Principle." In *The Portable Thoreau*, Carl Bode, ed. New York: Penguin, 1979, 631–655.

Thornton, James. "The State of the Environmentalists 1993." A Report to the Nathan Cummings Foundation and the Natural Resources Defense Council, 1993.

Thornton, James. "Radical Confidence: What is Missing from Eco-activism." *Tricycle*, vol. 3, no. 2, winter 1993.

Todd, Emmanuel. *The Explanation of Ideology*. New York: Blackwell, 1985.

Tuan, Yi-Fu. *Topohilia: A Study of Environmental Perception, Attitudes and Values*. Englewood Cliffs, N.J.: Praeger, 1974.

Turner, Frederick. *Beyond Geography: The Western Spirit Against the Wilderness*. New Brunswick, N.J.: Rutgers University Press, 1983.

Turner, Frederick. *Rediscovering America: John Muir in His Time and Ours*. San Francisco: Sierra Club, 1985.

Weinstein, Fred. *History and Theory After the Fall: An Essay on Interpretation*. Chicago: University of Chicago Press, 1990.

Weinstein, Fred, and Gerald Platt. *Psychoanalytic Sociology*. Baltimore: Johns Hopkins University Press, 1973.

Williams, Terry Tempest. *An Unspoken Hunger*. New York: Pantheon, 1994.

Wilson, Alexander. *The Culture of Nature: North American Landscape from Disney to the* Exxon Valdez. Cambridge, Mass.: Blackwell, 1992.

Wilson, Edward O. *Biophilia: The Human Bond with Other Species*. Cambridge, Mass.: Harvard University Press, 1984.

Wilson, Edward O. *The Diversity of Life*. Cambridge, Mass.: Harvard University Press, 1992.

Wilson, Edward O. "Biophilia and the Conservation Ethic." In *The Biophilia Hypothesis*. Stephen R. Kellert and Edward O. Wilson, eds. Washington, D.C.: Island Press, 1993.

Winner, Langdon. *The Whale and the Reactor: A Search for Limits in an Age of High Technology*. Chicago: University of Chicago Press, 1986.

Worster, Donald. *Nature's Economy: A History of Ecological Ideas*. New York: Cambridge University Press, 1985.

Young, Stanley. "An Interview with Thay." *The Mindfulness Bell 7*, summer/fall, 1992.

Index